Pre-Lab Exercises to accompany

EXPERIMENTAL ORGANIC CHEMISTRY

second edition

A Miniscale & Microscale Approach

John C. Gilbert
Stephen F. Martin
Royston M. Roberts†
The University of Texas at Austin

SAUNDERS GOLDEN SUNBURST SERIES
Saunders College Publishing
Harcourt Brace College Publishers

Fort Worth Philadelphia San Diego New York Orlando Austin
San Antonio Toronto Montreal London Sydney Tokyo

Although for mechanical reasons all pages of this publication are perforated, only those pages imprinted with Saunders College Publishing are intended for removal.

Requests for permission to make copies of any part of the work should be mailed to: Permissions Department, Harcourt Brace & Company, 6277 Sea Harbor Drive, Orlando, Florida 32887-6777.

Printed in the United States of America

Gilbert & Martin: Pre- Lab Exercises to accompany Experimental Organic Chemistry: A Miniscale and Microscale Approach, Second Edition.

0-03-024748-9

9012345 202 765432

Pre-Lab Exercises for RECRYSTALLIZATION. Part A. SOLVENT SELECTION (Section 3.2)

NAME (print): ______________________________ DATE: __________

INSTRUCTOR: ______________________________ LABORATORY SECTION: __________

1. By marking yes (Y) or no (N) in the space provided, specify which of the following criteria are met by a good solvent for a recrystallization.

    ______ **a.** The solutes are soluble in the cold solvent.

    ______ **b.** The solvent does not react chemically with the solutes.

    ______ **c.** The solvent is polar rather than nonpolar.

    ______ **d.** The boiling point of the solvent is above 100 °C.

    ______ **e.** The boiling point of the solvent preferably is below the melting point of the solute.

2. Give the criterion applied in this experimental procedure to classify a solute as being "soluble" in a particular solvent.

3. Review the functional groups present in resorcinol, benzoic acid, naphthalene, and acetanilide. Predict whether these molecules are expected to be polar (P) or nonpolar (NP).

    Resorcinol ______ Benzoic acid ______ Naphthalene ______ Acetanilide ______

4. Why is it important to

    a. avoid inhaling vapors of organic solvents?

    b. know the location and operating instructions of the nearest fire extinguisher when using 95% ethanol or petroleum ether for testing solubilities?

5. The flash point s of 95% ethanol and petroleum ether (bp 60–80 °C) are________ and ________, respectively.

6. Values of the oral LD_{50} (mg/kg) in rats of resorcinol, benzoic acid, naphthalene, and acetanilide are ________, ________, ________, and ________, respectively.

Pre-Lab Exercises for RECRYSTALLIZATION. Part B. RECRYSTALLIZING IMPURE SOLIDS (Section 3.2)

NAME (print): ______________________________ DATE: __________

INSTRUCTOR: ______________________________ LABORATORY SECTION: __________

1. By marking gravity (G) or vacuum (V) in the space provided, indicate which of the two different filtering techniques is more suitable for each of the following operations.

   _____ **a.** Hot filtration.

   _____ **b.** Removing decolorizing carbon.

   _____ **c.** Isolating recrystallized solute from solution.

2. Why is *flameless* heating used for heating a solution in hexane or diethyl ether during a recrystallization?

3. Why should the size of crystals obtained in a recrystallization be neither too large nor too small?

4. What is the process of seeding, as it applies to recrystallization? What purpose does it serve?

5. What is meant by the term, "oiling out," as it applies to crystallizations?

6. How is the purity of a recrystallized solid assessed?

7. Why should decolorizing carbon *not* be added to a solvent that is at or near its boiling point?

8. Why is it important to

a. break (terminate) the vacuum before turning off the water aspirator pump when employing the equipment for vacuum filtration shown in Figure 2.52?

b. avoid inhaling vapors of organic solvents?

c. know the position and operating instructions of the nearest fire extinguisher when using methanol, 95% ethanol, or 2-propanol as a crystallization solvent?

d. use a *fluted* filter paper for hot filtration?

Pre-Lab Exercises for RECRYSTALLIZATION. Part B. RECRYSTALLIZING IMPURE SOLIDS (Section 3.2)

NAME (print): ______________________________ DATE: __________

INSTRUCTOR: ______________________________ LABORATORY SECTION: __________

1. By marking gravity (G) or Craig tube (CT) in the space provided, indicate which of the two different filtering techniques is more suitable for each of the following operations.

   ______ **a.** Hot filtration.

   ______ **b.** Removing decolorizing carbon.

   ______ **c.** Isolating recrystallized solute from solution.

2. Why is *flameless* heating used for heating a solution in hexane or diethyl ether during a recrystallization?

3. Why should the size of crystals obtained in a recrystallization be neither too large nor too small?

4. What is the process of seeding, as it applies to recrystallization? What purpose does it serve?

5. What is meant by the term, "oiling out," as it applies to crystallizations?

6. How is the purity of a recrystallized solid assessed?

7. Why should decolorizing carbon *not* be added to a solvent that is at or near its boiling point?

8. Why is it important to

a. balance the weight when loading the centrifuge for a Craig tube filtration?

b. avoid inhaling vapors of organic solvents?

c. know the position and operating instructions of the nearest fire extinguisher when using methanol, 95% ethanol, or 2-propanol as a crystallization solvent?

d. use a *fluted* filter paper for hot filtration?

Pre-Lab Exercises for MELTING POINTS
(Section 3.3)

NAME (print): ____ DATE: ____________

INSTRUCTOR: ______________________________ LABORATORY SECTION: ________

1. List a convenient source of data concerning the physical constants and properties of organic compounds.

2. Indicate which of the following statements is true (T) and which is false (F).

______ **a.** An impurity raises the melting point of an organic compound.

______ **b.** A eutectic mixture has a sharp melting point, just as does a pure compound.

______ **c.** If the rate of heating of the oil bath used in a melting-point determination is too high, the melting point that results will likely be too low.

______ **d.** The sample should not be packed tightly into a capillary melting-point tube.

______ **e.** A heating bath containing mineral oil should not be used to determine the melting points of solids melting above 200 °C.

3. On the figure below, sketch the location of the capillary melting-point tube and the sample contained in it relative to the bulb of the thermometer.

4. What is the approximate rate at which the temperature of the heating bath should be increasing at the time the sample undergoes melting?

5. What is the preferred technique for accurately determining the melting point of an unknown compound in a minimum length of time?

6. How does measuring a mixed melting point help in determining the possible identity of two solid samples?

7. Briefly describe the technique for packing a capillary melting-point tube.

8. Why is it important to calibrate a thermometer with a set of standards having a *range* of melting points?

9. What toxic fumes are evolved by burning mineral oil?

NAME (print): ______________________________ DATE: __________

INSTRUCTOR: ______________________________ LABORATORY SECTION: __________

Questions 1, 2, and 4 may be answered by marking yes (Y) or no (N) in the space provided for each part.

1. The addition of a nonvolatile solute to a volatile liquid

______ **a.** has no effect on the boiling point of the volatile liquid.

______ **b.** lowers the boiling point of the volatile liquid.

______ **c.** raises the boiling point of the volatile liquid.

2. The boiling point of a pure liquid

______ **a.** is the same in Denver, Colorado (elevation 5280 ft) as it is in San Francisco, California.

______ **b.** is lower in Denver than in San Francisco.

______ **c.** is usually found to be almost the same as the reported "standard boiling temperature" in San Francisco.

3. Explain your answers to Exercise 2.

4. The boiling point, as determined in the miniscale boiling-point apparatus, is the temperature

______ **a.** of the liquid at the time bubbles first emerge slowly from the liquid.

______ **b.** at the vapor-liquid interface above the surface of the boiling liquid while a drop of liquid is suspended from the thermometer.

______ **c.** of the liquid at the time bubbles emerge rapidly from the liquid.

______ **d.** of the heating source at the time bubbles emerge rapidly from the liquid.

5. Why is heating a liquid in a closed system dangerous?

6. What precautions should be taken *before* lighting a Bunsen burner or microburner in the laboratory?

7. Why is mineral oil an inappropriate heating fluid for determining the boiling points of samples that exceed 200 °C?

NAME (print): ______________________________ DATE: __________

INSTRUCTOR: ______________________________ LABORATORY SECTION: __________

Questions 1, 2, and 4 may be answered by marking yes (Y) or no (N) in the space provided for each part.

1. The addition of a nonvolatile solute to a volatile liquid

______ **a.** has no effect on the boiling point of the volatile liquid.

______ **b.** lowers the boiling point of the volatile liquid.

______ **c.** raises the boiling point of the volatile liquid.

2. The boiling point of a pure liquid

______ **a.** is the same in Denver, Colorado (elevation 5280 ft) as it is in San Francisco, California.

______ **b.** is lower in Denver than in San Francisco.

______ **c.** is usually found to be almost the same as the reported "standard boiling temperature" in San Francisco.

3. Explain your answers to Exercise 2.

4. The boiling point, as determined in the micro boiling-point apparatus, is the temperature at the time

______ **a.** bubbles first emerge slowly from the inverted capillary tube.

______ **b.** bubbles begin to emerge rapidly from the inverted capillary tube.

______ **c.** the liquid begins to re-enter and rise in the inverted capillary tube.

5. What are the bubbles that emerge slowly from the inverted capillary tube before the boiling temperature is reached, and why does this occur?

6. Define a *closed system* as it applies to a Thiele tube and the liquids it contains.

7. Why should a closed system, as defined in exercise 6, *not* be heated unless special apparatus is available?

8. Why is the micro boiling-point apparatus used in this procedure *not* a closed system?

9. Why is mineral oil an inappropriate heating fluid for determining the boiling points of samples that exceed 200 °C?

10. What precautions should be taken *before* lighting a Bunsen burner or microburner in the laboratory?

NAME (print): ______________________________ DATE: __________

INSTRUCTOR: ______________________________ LABORATORY SECTION: __________

1. By marking simple (S) or fractional (F) in the space provided, indicate which of the two distillation techniques would be more suitable for the following.

   ______ **a.** preparing drinking water from sea water.

   ______ **b.** removing diethyl ether, bp 35 °C (760 Torr), from a solution containing *p*-dichlorobenzene, bp 174 °C (760 torr).

   ______ **c.** separating benzene, bp 80 °C (760 Torr), from toluene, bp 111 °C (760 Torr).

2. What is the purpose of the stirbar placed in the stillpot?

3. How does the composition of the liquid at the top of a fractional distillation column compare with the composition of the liquid at the bottom of a column? (Answer in terms of the relative amounts of lower-boiling and higher-boiling components.)

4. Two fractionating columns are each 40 cm in length. Column A has HETP = 2 cm and column B has HETP = 20 cm. By marking in the space provided, indicate whether column A or B would be more suitable to separate a binary mixture in which the components differ in boiling point by 10 °C?

   ______

5. Define the term, *reflux ratio*.

6. Why is it important to align the fractionating column as nearly vertical as possible?

7. On the figure below, sketch the correct location of the thermometer bulb during a miniscale distillation.

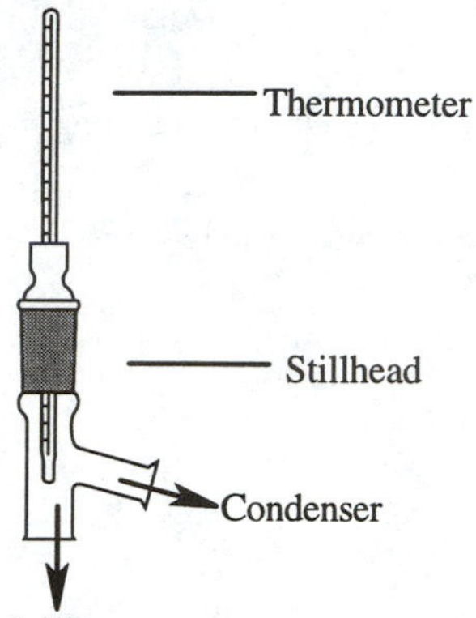

8. Why is it important that a drop of condensate be suspended from the thermometer during a distillation.

9. With respect to the condenser used in an apparatus for simple or fractional distillation, why should the lower rather than the upper nipple be used for the water inlet?

10. The flash point (°C) of cyclohexane is _______; that of toluene is _______.

11. List the target organs affected by cyclohexane and toluene.

NAME (print): ______________________________ DATE: __________

INSTRUCTOR: ______________________________ LABORATORY SECTION: __________

1. Indicate whether each of the following statements is true (T) or false (F).

_____ **a.** Simple distillation may be used to prepare drinking water from sea water.

_____ **b.** Simple distillation may be used to remove diethyl ether, bp 35 °C (760 Torr), from *p*-dichlorobenzene, bp 174 °C (760 torr).

_____ **c.** Simple distillation may be used to separate benzene, bp 80 °C (760 Torr), from toluene, bp 111 °C (760 Torr).

2. What is the purpose of the spinvane placed in the stillpot?

3. On the figure below, sketch the correct location of the thermometer bulb during a microscale, simple distillation.

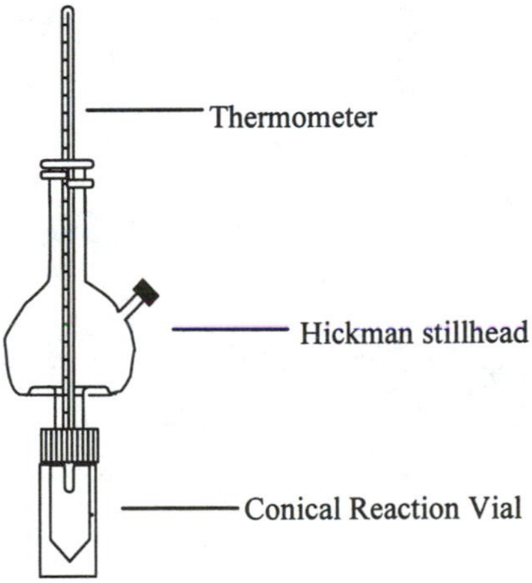

4. Why is it important that a drop of condensate be suspended from the thermometer during a distillation.

5. With respect to the condenser used in an apparatus for simple distillation, why should the lower rather than the upper nipple be used for the water inlet?

6. The flash point (°C) of cyclohexane is________.

7. List the target organs affected by cyclohexane.

Pre-Lab Exercises for STEAM DISTILLATION OF CITRAL FROM LEMON GRASS OIL (Section 4.5)

NAME (print): ______________________ DATE: __________

INSTRUCTOR: ______________________ LABORATORY SECTION: __________

Questions 1 and 3 may be answered by marking (T) or (F) in the space provided for each part to indicate whether the statement is true or false.

1. Steam distillation would be the procedure of choice for separating mixtures of

______ **a.** methanol, bp 65 °C (760 Torr), and water; methanol is completely miscible with water.

______ **b.** *p*-dichlorobenzene, bp 174 °C (760 Torr), and water; *p*-dichlorobenzene is insoluble in water.

______ **c.** Ethylene glycol ($HOCH_2CH_2OH$), bp 196 °C (760 Torr), and water; the glycol is miscible with water in all proportions.

2. Explain your answer to each part of Exercise 1.

a.

b.

c.

3. Indicate whether the following statement is true (T) or false (F): The boiling point during a steam distillation is always less than 100 °C.

______

4. Explain your answer to Exercise 3.

5. In what way are the advantages of steam distillation and vacuum distillation similar?

6. What kind of mixture may better be separated by steam distillation than by vacuum distillation?

7. Write structural formulas for the two diastereomers (geometrical isomers) of citral.

8. Based on a consideration of the nature of the groups present in the diastereomers that comprise citral, why is citral insoluble in water?

9. Why is a flame *not* used to assist in the removal of diethyl ether from citral?

10. Why should ethereal solutions *not* be stored in your laboratory locker from one period to the next?

NAME (print): ______________________________ DATE: ____________

INSTRUCTOR: ______________________________ LABORATORY SECTION: __________

1. Write balanced equations for all chemical reactions that should occur in the extraction procedure you are to perform.

2. Devise a flow chart for purification that summarizes the separations to be performed in this experiment. Consult Figure 1.2 for an example of such a flow chart.

3. When extracting an aqueous solution with an organic solvent, if you are uncertain as to which layer is aqueous, how could you settle the issue?

4. Which layer, upper (U) or lower (L), will the following solvents usually form when used to extract dilute aqueous solutions: diethyl ether ________, dichloromethane ________, chloroform________, hexane ________?

5. Calculate the amount of acid needed to neutralize the basic extract in the experiment you are to perform. Show your work.

6. Indicate which of the following statements is true (T) and which is false (F).

______ **a.** Benzoic acid forms a water-soluble salt, whereas naphthalene does not.

______ **b.** Carboxylic acids containing six or more carbon atoms per molecule are more soluble in diethyl ether than in water.

______ **c.** Carboxylic acids containing six or more carbon atoms per molecule are more soluble in 2.5 *M* sodium hydroxide than in diethyl ether.

______ **d.** Naphthalene is more soluble in diethyl ether than is sodium benzoate.

______ **e.** As a general rule, aqueous sodium bicarbonate is preferred to aqueous sodium hydroxide for abstracting acidic compounds from organic solutions.

______ **f.** A criterion for a dry organic solution is that the solution is not cloudy.

______ **g.** Drying agents need not be removed prior to removing solvents when isolating products.

7. Values of the oral LD_{50} (mg/kg) in rats of benzoic acid and naphthalene are________ and________, respectively.

8. Why is frequent venting necessary when extracting the dichloromethane solution with aqueous base?

NAME (print): ______________________________ DATE: __________

INSTRUCTOR: ______________________________ LABORATORY SECTION: __________

1. Write balanced equations for all chemical reactions that should occur in the extraction procedure you are to perform.

2. Devise a flow chart for purification that summarizes the separations to be performed in this experiment. Consult Figure 1.2 for an example of such a flow chart.

3. When extracting an aqueous solution with an organic solvent, if you are uncertain as to which layer is aqueous, how could you settle the issue?

4. Which layer, upper (U) or lower (L), will the following solvents usually form when used to extract dilute aqueous solutions: diethyl ether ________, dichloromethane ________, chloroform________, hexane ________?

5. Calculate the amount of acid needed to neutralize the basic extracts obtained in this experiment. Show your work.

6. Indicate which of the following statements is true (T) and which is false (F).

______ a. Benzoic acid forms a water-soluble salt, whereas naphthalene does not.

______ b. Carboxylic acids containing six or more carbon atoms per molecule are more soluble in diethyl ether than in water.

______ c. Phenols containing six or more carbon atoms per molecule are more soluble in 2.5 *M* sodium hydroxide than in diethyl ether.

______ d. Naphthalene is more soluble in diethyl ether than is sodium phenoxide.

______ e. As a general rule, aqueous sodium bicarbonate is preferred to aqueous sodium hydroxide for abstracting acidic compounds from organic solutions.

______ f. A criterion for a dry organic solution is that the solution is not cloudy.

______ g. Drying agents need not be removed prior to removing solvents when isolating products.

7. Values of the oral LD_{50} (mg/kg) in rats of benzoic acid, phenol, and naphthalene are ________, ________, and ________ respectively.

8. Why is frequent venting necessary when extracting the dichloromethane solution with aqueous base?

NAME (print): ______________________________ DATE: __________

INSTRUCTOR: ______________________________ LABORATORY SECTION: __________

1. The technique responsible for isolation of trimyristin in this experiment is an example of (check one)

______ liquid-liquid, ______ solid-liquid, ______ gas-liquid partitioning.

2. Indicate which of the following statements is true (T) and which is false (F).

______ **a.** Pure trimyristin is a liquid at room temperature.

______ **b.** According to the equation defining the distribution coefficient *K*, a value of 2 for *K* means that *A* is more soluble in solvent S_o than in solvent S_x

$$K = \frac{grams\ of\ A\ in\ S_x}{grams\ of\ A\ in\ S_o} x \frac{mL\ of\ S_o}{mL\ of\ S_x}$$

______ **c.** Nutmeg contains a complex mixture of organic compounds that are soluble in diethyl ether.

______ **d.** Trimyristin is more soluble in acetone than in diethyl ether.

______ **e.** The functional group present in trimyristin is a carboxylic acid group.

______ **f.** The term *lipid* describes a category of organic compounds that are insoluble in water.

______ **g.** With respect to the technique of extraction, the term *partitioning* means physical separation of two immiscible phases by an impermeable membrane.

3. Explain your answer to Exercises 2e and 2f.

e.

f.

4. Why is trimyristin considered a saturated fat?

5. What is wrong with the experimental set-up shown below for extraction of nutmeg?

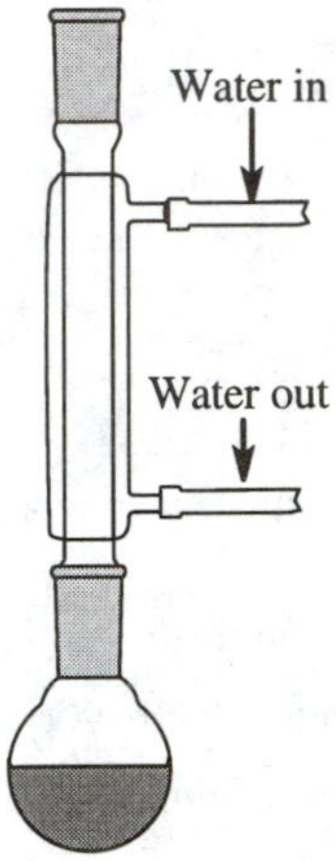

6. On the figure shown in Exercise 5, indicate on the condenser the *upper* limit for the ring of condensate and the points at which clamps should be located.

7. Why is a flame *not* to be used to heat diethyl ether at reflux in this experiment?

8. What would the consequence be of not having the condenser tightly connected to the round-bottom flask during the reflux period?

9. The flash point (°C) of diethyl ether is ________; that of acetone is ________.

10. The concentration, in ppm, at which diethyl ether causes human eye irritation is ________ and that for acetone is ________.

NAME (print): ______________________________ DATE: __________

INSTRUCTOR: ______________________________ LABORATORY SECTION: __________

1. The technique responsible for isolation of trimyristin in this experiment is an example of (check one) ______ liquid-liquid, _____ solid-liquid, ______ gas-liquid extraction.

2. Indicate which of the following statements is true (T) and which is false (F).

______ **a.** Pure trimyristin is a liquid at room temperature.

______ **b.** According to the equation defining the distribution coefficient *K*, a value of 2 for *K* means that *A* is more soluble in solvent S_o than in solvent S_x

$$K = \frac{grams\ of\ A\ in\ S_x}{grams\ of\ A\ in\ S_o} x \frac{mL\ of\ S_o}{mL\ of\ S_x}$$

______ **c.** Nutmeg contains a complex mixture of organic compounds that are soluble in diethyl ether.

______ **d.** Trimyristin is more soluble in acetone than in diethyl ether.

______ **e.** The functional group present in trimyristin is a carboxylic acid group.

______ **f.** The term *lipid* describes a category of organic compounds that are insoluble in water.

______ **g.** With respect to the technique of extraction, the term *partitioning* means physical separation of two immiscible phases by an impermeable membrane.

3. Explain your answer to Exercises 2e and 2f.

e.

f.

4. Why is trimyristin considered a saturated fat?

5. What is wrong with the experimental set-up shown below for extraction of nutmeg?

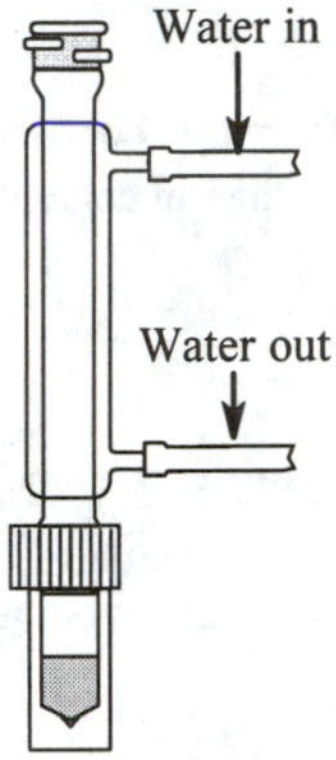

6. On the figure shown in Exercise 5, indicate on the condenser the *upper* limit for the ring of condensate and the point at which a clamp should be located.

7. Why is a flame *not* to be used to heat diethyl ether at reflux in this experiment?

8. What would the consequence be of not having the condenser tightly connected to the conical vial during the reflux period?

9. The flash point (°C) of diethyl ether is ________; that of acetone is ________.

10. The concentration, in ppm, at which diethyl ether causes human eye irritation is ________ and that for acetone is ________.

NAME (print): ______________________________ DATE: __________

INSTRUCTOR: __________________________ LABORATORY SECTION: ________

1. What difficulty may result if

a. a chromatographic column is not placed in a vertical position?

b. the liquid level of the eluent is allowed to drop below the top of the column?

2. Petroleum ether followed by dichloromethane is used to separate fluorene and fluorenone. Why would reversing the order in which these solvents are used be unwise?

3. What is the best experimental procedure to use in choosing an eluting solvent for column chromatography?

4. Define:

a. eluant

b. eluate

c. adsorption

5. Which type of alumina, "acidic" or "basic," would provide for the better separation of acids?

6. How can the adsorptivity (activity) of alumina be varied?

7. Why should a mixture to be separated be introduced onto a column in a minimum amount of solvent?

8. Why should no flames be allowed on the lab bench when performing this experiment?

9. Underline the media that are appropriate for extinguishing fires involving fluorene or fluorenone:

   Water Carbon dioxide Chemical powder Foam

10. With what general class of chemical reagents are fluorene and fluorenone incompatible?

Pre-Lab Exercises for THIN LAYER CHROMATOGRAPHY. Part A. SEPARATION OF SPINACH PIGMENTS BY TLC (Section 6.3)

NAME (print): ______________________________ DATE: __________

INSTRUCTOR: ______________________________ LABORATORY SECTION: __________

1. Describe three ways in which colorless compounds can be located on a TLC plate.

2. Check TLC (thin-layer chromatography) or CC (column chromatography) as the more appropriate answer to the following questions or statements.

a. TLC ________ CC ________ is a quicker procedure for separating components of a mixture.

b. In TLC ________ CC ________ the solvent front moves downward.

c. TLC ________ CC ________ is better for separating a 5-gram mixture of components.

d. TLC ________ CC ________ is better for separating a mixture of volatile compounds.

3. Why should a TLC plate be removed from the solvent *before* the solvent front reaches the top of the plate?

4. What is the purpose of shaking the petroleum ether-ethanol extract of the leaves with water in a separatory funnel?

5. The separating power, or activity, of a TLC plate is increased by heating the plate in an oven at 100 °C. Why? (*Hint:* See Section 6.2)

6. Why should the developing chamber for a TLC plate *not* be open to the atmosphere?

7. Which of the following diagrams illustrate(s) an *improper* way of spotting a TLC plate? Tell what is wrong in each case.

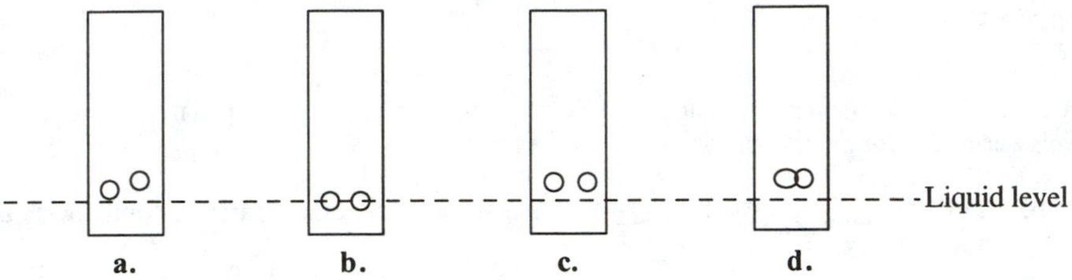

a.

b.

c.

d.

Pre-Lab Exercises for THIN LAYER CHROMATOGRAPHY. Part B. SEPARATION OF *SYN*- AND *ANTI*-AZOBENZENES BY TLC (Section 6.3)

NAME (print): ________________________________ DATE: ____________

INSTRUCTOR: ________________________________ LABORATORY SECTION: __________

1. Describe three ways in which colorless compounds can be located on a TLC plate.

2. Check TLC (thin-layer chromatography) or CC (column chromatography) as the more appropriate answer to the following questions or statements.

a. TLC ________ CC ________ is a quicker procedure for separating components of a mixture.

b. In TLC ________ CC ________ the solvent front moves downward.

c. TLC ________ CC ________ is better for separating a 5-gram mixture of components.

d. TLC ________ CC ________ is better for separating a mixture of volatile compounds.

3. Why should a TLC plate be removed from the solvent *before* the solvent front reaches the top of the plate?

4. The separating power, or activity, of a TLC plate is increased by heating the plate in an oven at 100 °C. Why? (*Hint:* See Section 6.2)

5. Why should the developing chamber for a TLC plate *not* be open to the atmosphere?

6. Which of the following diagrams illustrate(s) an *improper* way of spotting a TLC plate? Tell what is wrong in each case.

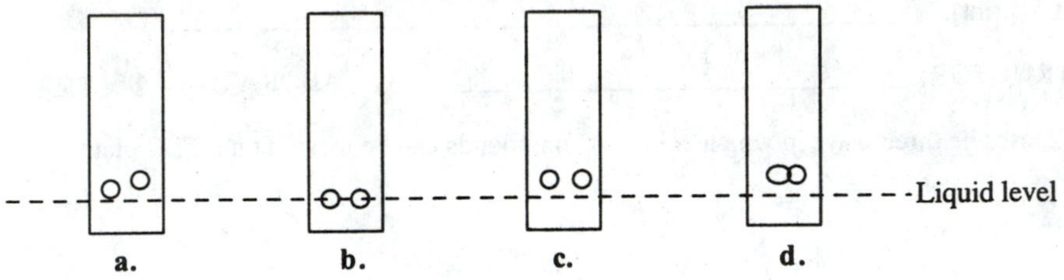

a.

b.

c.

d.

7. What toxic fumes are evolved in fires involving azobenzene?

8. The oral toxicity (g/kg) of azobenzene in rats is ________.

Pre-Lab Exercises for GAS-LIQUID CHROMATOGRAPHY. PART A. QUALITATIVE AND QUANTITATIVE ANALYSIS OF A MIXTURE BY GLC (Section 6.4)

NAME (print): ______________________________ DATE: __________

INSTRUCTOR: ______________________ LABORATORY SECTION: _________

1. Explain how *liquids* can be analyzed by gas chromatography.

2. Briefly define or describe the function, in gas chromatography, of the

a. carrier gas

b. stationary liquid phase

c. solid support

3. Which column material should be chosen to separate a mixture of

	Carbowax 20M	SE 30
a. alcohols	_____	_____
b. aromatic hydrocarbons	_____	_____

4. If ethyl acetate and *n*-butyl acetate are analyzed by gas chromatography, which of these esters will generally produce a peak with the *shorter* retention time?

5. How could you confirm your answer to Exercise 4 by experiment?

6. What operating variables determine retention time?

7. Explain why is it important to inject the sample quickly?

8. The flash points (°C) of ethylbenzene, toluene, and isopropylbenzene are ________, ________, and ________, respectively.

9. Underline the media that are appropriate for extinguishing fires involving the hydrocarbons listed in Exercise 8: Water Carbon dioxide Chemical powder Foam

Pre-Lab Exercises for GAS-LIQUID CHROMATOGRAPHY. PART B. DETERMINING GLC RESPONSE FACTORS FOR KNOWN COMPOUNDS (Section 6.4)

NAME (print): ______________________________ DATE: __________

INSTRUCTOR: ______________________________ LABORATORY SECTION: __________

1. Explain why it is important to mix the solutions containing two components completely.

2. Typical standards include ethyl acetate, *p*-xylene and ethanol. The flash point (°C) of ethyl acetate is ________, of *p*-xylene is ________ and of ethanol is ________.

3. Explain why is it important to inject the sample quickly?

4. Indicate whether the following statement is true (T) or false (F): It is acceptable for the standard and the sample whose response factor is being determined to have the same retention time under the chromatographic conditions of the measurement. _______

5. Explain your answer to Exercise 4.

6. Indicate whether the following statement is true (T) or false (F): In order to obtain accurate values for the GLC response factor of the unknown sample, it is *essential* to use *equal* volumes of the standard and the sample whose response factor is being determined.

____

7. Explain your answer to Exercise 6.

8. Indicate whether the following statement is true (T) or false (F): To minimize the error in determining the GLC response factor of an unknown, the peaks on the chromatogram for both the standard and the unknown should be as large as possible.

____

9. Explain your answer to Exercise 8.

Pre-Lab Exercises for SEPARATION OF DIASTEREOMERIC 1,2-CYCLO-HEXANEDIOLS (SECTION 7.2)

NAME (print): ______________________________ DATE: ____________

INSTRUCTOR: ______________________________ LABORATORY SECTION: ____________

1. Indicate which of the following statements is true (T) and which is false (F).

_______ **a.** *Trans*-1,2-cyclohexanediol is a *meso* form.

_______ **b.** The *cis*- and *trans*-1,2-cyclohexanediols are enantiomers.

_______ **c.** The *cis*- and *trans*-1,2-cyclohexanediols cannot be separated by fractional crystallization.

_______ **d.** (+)-*Trans*-Cyclohexanediol and (-)-*trans*-cyclohexane-1,2-diol have different chromatographic adsorption properties.

2. Why should a TLC plate be removed from the solvent before the solvent front reaches the top of the plate?

3. Why should the developing chamber for a TLC plate not be open to the atmosphere?

4. What physical properties could be used to distinguish *cis*- from *trans*-1,2-cyclohexanediol?

5. Why might water *not* be an appropriate extinguishing medium for burning petroleum ether?

6. Identify any of the following diagrams that illustrate an *improperly* spotted TLC plate and explain what is wrong in each such case.

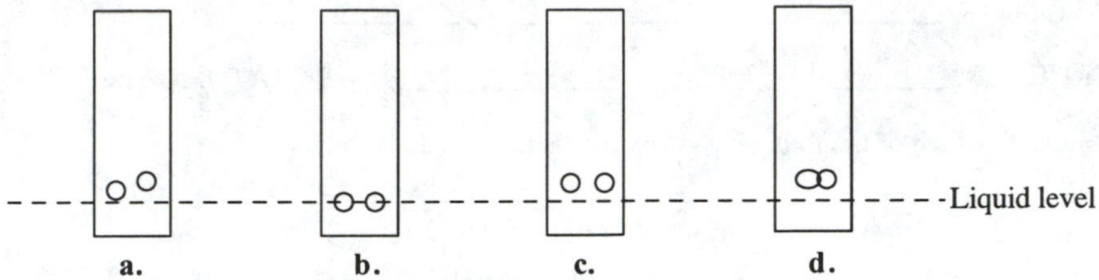

a.

b.

c.

d.

7. The flash points (°C) for petroleum ether (bp 60–80 °C), acetone, and 2-propanol are, respectively, ________, ________, and ________.

NAME (print): ______________________________ DATE: __________

INSTRUCTOR: ______________________________ LABORATORY SECTION: __________

1. Write structural formulas for dimethyl maleate and dimethyl fumarate.

2. Indicate which of the following statements is true (T) and which is false (F).

______ **a.** Dimethyl maleate and dimethyl fumarate are enantiomers.

______ **b.** Dimethyl maleate and dimethyl fumarate are diastereomers.

______ **c.** Dimethyl maleate and dimethyl fumarate are conformational isomers.

______ **d.** Dimethyl maleate and dimethyl fumarate are constitutional isomers.

3. Put a check mark beside the ester that

a. has the higher melting point, dimethyl maleate ______ dimethyl fumarate ______?

b. has the higher boiling point, dimethyl maleate ______ dimethyl fumarate ______?

c. is more soluble in CH_2Cl_2, dimethyl maleate ______ dimethyl fumarate ______?

4. Why should glassware containing residues of bromine *not* be rinsed with acetone?

5. State the recommended procedure in this experiment to be used for destroying residual bromine and write the equation for the reaction that occurs.

6. What is to be done if the melting point of the isolated dimethyl fumarate is lower than the reported melting point of this compound?

7. What is the vapor pressure of bromine at 20 °C?

8. Why should inhalation of the vapors of bromine be avoided?

9. What should you do if any bromine comes in contact with your skin?

10. Provide the flash points (°C) of dichloromethane _______ and of ethanol _______.

Pre-Lab Exercises for PROPERTIES OF THE ENANTIOMERS OF CARVONE (Section 7.4)

NAME (print): ______________________________ DATE: __________

INSTRUCTOR: ______________________________ LABORATORY SECTION: __________

1. Put a check mark beside any of the following physical properties that would be expected to be the same for the enantiomeric carvones.

   boiling point ______, solubility in acetone ______, the R_f value in TLC ______, odor ______,

   retention time in GLC ______, rotation of plane-polarized light ______.

2. Indicate which of the following statements is true (T) and which is false (F).

   ______ a. (*R*)-(-)-Carvone and limonene are diastereomers.

   ______ b. (*R*)-(-)-Carvone and (*S*)-(+)-carvone are both volatile liquids.

   ______ c. (*R*)-(-)-Carvone and limonene are not isomers of one another.

   ______ d. (*R*)-(-)-Carvone and limonene are constitutional isomers.

3. What color change is expected when a sample of a carvone is treated with Br_2 in CH_2Cl_2?

4. What color change is expected when a sample of a carvone is treated with aqueous $KMnO_4$?

5. Answer the previous two questions for the case in which limonene rather than a carvone is used.

6. What do you expect to observe when a carvone is treated with 2,4-dinitrophenylhydrazine?

7. What is the vapor pressure of bromine at 20 °C?

8. Why should inhalation of the vapors of bromine be avoided?

9. What should you do if any bromine comes in contact with your skin?

10. The flash points (°C) of ethanol, ethyl acetate, and dichloromethane are ________, ________, and ________, respectively.

11. What may occur if 2,4-dinitrophenylhydrazine is absorbed into the body?

NAME (print): ________________________________ DATE: ____________

INSTRUCTOR: ____________________________ LABORATORY SECTION: __________

1. Put a check mark beside any of the following physical properties that would be expected to be the same for the enantiomeric 1-phenylethanamines.

boiling point ______, solubility in acetone ______, the R_f value in TLC ______, odor ______,

retention time in GLC ______, rotation of plane-polarized light ______.

2. Methanol is to be used as the solvent for the separation of the diastereomeric forms of 1-phenylethanamine hydrogen tartrate. Briefly explain why this solvent is better for the separation than

a. petroleum ether (bp 60–80 °C/760 Torr).

b. water.

3. Why is the solution of 1-phenylethanamine hydrogen tartrate not to be cooled in *ice water* before the crystals are collected?

4. Why cannot racemic tartaric acid be used to resolve 1-phenylethanamine?

5. Why cannot *meso* tartaric acid be used to resolve 1-phenylethanamine?

6. What is the crystalline form of the amine hydrogen tartrate to be isolated in this experiment?

7. What is the principle on which the resolution of the racemic 1-phenylethanamines by formation of a tartrate salt depends?

8. Why is *flameless* heating specified for preparing the methanolic solution of tartaric acid in this experiment?

9. The flash points (°C) of diethyl ether, methanol, and 1-phenylethanamine are ________, ________, and ________, respectively.

10. What toxicology data, if any, are provided in the MSDS sheets for tartaric acid?

NAME (print): ______________________ DATE: __________

INSTRUCTOR: ______________________ LABORATORY SECTION: __________

1. Write the structure of the substance to be used in this experiment to initiate free radical chain chlorination, and give the balanced equation for its thermal decomposition.

2. What is the molar ratio of initiator to sulfuryl chloride to be used in the procedure? Why is so little initiator needed relative to the amount of chlorinating agent used?

3. Write the balanced equation for the reaction of sulfuryl chloride with water.

4. Why would water be an inappropriate medium for extinguishing a fire involving sulfuryl chloride?

5. Why is a "gas trap" to be used in the chlorinatic . reaction?

6. What is the theoretical amount of weight that should be lost by the reaction mixture in the procedure you are to perform? Show your calculations.

7. Why should the separatory funnel be vented frequently when the reaction mixture is shaken with aqueous sodium carbonate? Write an equation for any chemical reaction that is responsible for the need for venting.

8. What type of distillation apparatus, fractional or simple, is to be used for isolation of the chlorinated product(s)? Why is this type selected?

9. Why are pieces of glass or of stainless steel "sponge" rather than copper "sponge" recommended as packing material for the distillation column?

10. What visual criterion is to be used to assess whether the reaction mixture is dry prior to distillation?

11. Put a check mark beside any of the following materials that evolves toxic fumes upon heating or burning: 1-chlorobutane______, sulfuryl chloride______, sodium chloride______, and sodium sulfate ______.

12. The oral LDL_O (mg/kg) in humans of bromine is________.

Pre-Lab Exercises for CHLORINATION USING SULFURYL CHLORIDE (Section 9.2)

NAME (print): ______________________________ DATE: ____________

INSTRUCTOR: ______________________________ LABORATORY SECTION: __________

1. Write the structure of the substance to be used in this experiment to initiate free radical chain chlorination, and give the balanced equation for its thermal decomposition.

2. What is the molar ratio of initiator to sulfuryl chloride to be used in the procedure? Why is so little initiator needed relative to the amount of chlorinating agent used?

3. Write the balanced equation for the reaction of sulfuryl chloride with water.

4. Why would water be an inappropriate medium for extinguishing a fire involving sulfuryl chloride?

5. Why is a "gas trap" to be used in the chlorination reaction?

6. What is the theoretical amount of weight that should be lost by the reaction mixture in the procedure you are to perform? Show your calculations.

7. Why should the conical vial be vented frequently when the reaction mixture is shaken with aqueous sodium carbonate? Write an equation for any chemical reaction that is responsible for the need for venting.

8. What technique is to be used to hasten the drying of the reaction mixture prior to distillation?

9. What visual criterion is to be used to assess whether the reaction mixture is dry prior to distillation?

10. Put a check mark beside any of the following materials that evolves toxic fumes upon heating or burning: 1-chlorobutane______, sulfuryl chloride______, sodium chloride______, and sodium sulfate______.

11. The oral LDL_O (mg/kg) in humans of bromine is________.

Pre-Lab Exercises for BROMINATION: SELECTIVITY OF HYDROGEN ATOM ABSTRACTION (Section 9.3)

NAME (print): ______________________________ DATE: __________

INSTRUCTOR: ______________________________ LABORATORY SECTION: __________

1. Draw a structure containing the specified type of hydrogen atom and circle the atom.

a. 1° aliphatic hydrogen atom.

b. 3° benzylic hydrogen atom.

c. vinylic hydrogen atom.

2. What experimental criterion is to be used to measure the rates of bromination of the hydrocarbons used in this experiment?

3. Underline the proper response in the following:

a. The results of this experiment will allow determination of the (relative, absolute) rates of bromination of a series of hydrocarbons.

b. The mechanism of the bromination is classified as a(n) (electrophilic addition, nucleophilic substitution, free-radical substitution) process.

c. Molecular bromine is a (solid, liquid, gas) at room temperature and is (corrosive, non-corrosive) to the skin.

4. Why should apparatus containing residues of bromine *not* be rinsed with acetone? How can such residues be chemically removed?

5. Determine the average ratio of Br_2 to hydrocarbon to be used in this experiment. To perform the calculation, assume that the hydrocarbon has a density of 0.8 and a molecular weight of 100. Show your work.

6. Why are the hydrocarbons to be used in excess in this experiment?

7. Why is the bromine that is to be added to each test tube in this experiment measured out as a solution of bromine in dichloromethane rather than as pure bromine?

8. The oral LDL_o (mg/kg) in humans of bromine is ________ and of toluene is ________.

9. The flash point (°C) of toluene is ________; that of methylcyclohexane is ________.

10. Underline the media that are appropriate for extinguishing fires involving toluene, *tert*-butylbenzene and ethylbenzene: Water Carbon dioxide Chemical powder Foam

NAME (print): ______________________________ DATE: ____________

INSTRUCTOR: ______________________ LABORATORY SECTION: __________

1. Calculate the molar ratio of base to 2-bromo-2-methylbutane to be used in the elimination experiment(s) you perform and specify which is "limiting reagent." Show your calculations.

2. Why is a Hempel column rather than a regular condenser to be used during the period of reflux?

3. Why is the column to be filled with a packing material during this stage of the procedure?

4. Why is the hold-up of the Hempel column when being used as a fractionating column greater when the column is packed rather than unpacked?

5. Why is the vacuum adapter of the apparatus to be fitted with a drying tube throughout the course of the reaction and the distillation?

6. Why is the receiving flask to be cooled in an ice-water bath throughout the reaction and distillation?

7. Why is it particularly important that the ground-glass joint linking the Hempel column to the reaction flask be properly lubricated for the base-promoted elimination of 2-methyl-2-bromobutane?

8. Write equations for the chemical reactions that you will use to demonstrate the presence of alkenes in your distilled product.

9. What is an appropriate extinguishing medium for fires involving potassium *tert*-butoxide?

10. The flash points (°C) of 2-methyl-1-butene and of 2-methyl-2-butene are ________ and ________, respectively.

11. The oral LDL_0 (mg/kg) in humans of bromine is ________.

NAME (print): ______________________ DATE: __________

INSTRUCTOR: ______________________ LABORATORY SECTION: __________

1. Calculate the molar ratio of potassium hydroxide to 2-bromo-2-methylbutane to be used in the experiment you perform and specify which is "limiting reagent." Show your calculations.

2. Why is a water-cooled rather than an air-cooled condenser specified in this procedure?

3. Why is an aqueous solution of potassium hydroxide an *unsuitable* medium for effecting the base-promoted elimination of this procedure?

4. What would the consequence be of *not* having the condenser tightly mated with the Hickman stillhead?

5. Why is necessary to stir the reaction mixture during the elimination reaction?

6. Why should the distillate be kept cold after it is removed from the Hickman stillhead?

7. Why is it particularly important that the ground-glass joint linking the Hickman stillhead to the conical vial be properly lubricated for the base-promoted elimination of 2-methyl-2-bromobutane?

8. Write equations for the chemical reactions that you will use to demonstrate the presence of alkenes in your distilled product.

9. What is an appropriate extinguishing medium for fires involving potassium *tert*-butoxide?

10. The flash points (°C) of 2-methyl-1-butene and of 2-methyl-2-butene are ________ and ________, respectively.

11. The oral LDL_0 (mg/kg) in humans of bromine is ________.

NAME (print): ______________________________ DATE: __________

INSTRUCTOR: ______________________________ LABORATORY SECTION: __________

1. What is the function of the acid catalyst in promoting the dehydration of alcohols?

2. Why would concentrated hydrochloric acid be an *inappropriate* catalyst for the dehydration of alcohols?

3. Why is the formation of substitution products involving displacement of water by attack of bisulfate upon a protonated alcohol *not* a reaction of concern in the elimination reaction?

4. Why is there an upper limit to the temperature at which the alkene(s) is (are) to be collected?

5. Write equations for the chemical reaction(s) that you will use to demonstrate the presence of alkene(s) in your distilled product.

6. Why is it important to dry the crude alkene(s) prior to the final distillation?

7. What visual criterion is to be used to assess whether the reaction mixture is dry prior to distillation?

8. Specify the "limiting reagent" in the dehydration procedure you perform. Give your reasoning.

9. Underline the media appropriate for extinguishing fires involving cyclohexene or the isomeric methylpentenes formed by dehydration of the corresponding alcohols: Water Carbon dioxide Chemical powder Foam

10. Specify whether the alkene or alkenes formed by the dehydration procedure you are to perform has a flash point above 25 °C? If so, give the flash point.

11. The oral LDL_0 (mg/kg) in humans of bromine is ________.

NAME (print): ______________________________ DATE: __________

INSTRUCTOR: ______________________________ LABORATORY SECTION: __________

1. Why do the conditions of this experiment favor electrophilic rather than free-radical addition of HBr to 1-hexene?

2. What is the role of the quaternary ammonium salt to be used in this experiment?

3. When concentrated aqueous HBr is added to 1-hexene, do you expect to observe a homogeneous or a heterogeneous reaction mixture? Why?

4. Why is vigorous stirring or other agitation of the reaction mixture important?

5. Write equations for the chemical reaction(s) responsible for the pressure build-up when the crude reaction mixture is washed with aqueous sodium bicarbonate.

6. If you are uncertain as to which layer is which when extracting an organic solution with an aqueous solution, how might you settle the issue?

7. Write balanced equations for the chemical reactions that could be used to demonstrate that 2- rather than 1-bromohexane has been formed in this experiment.

8. What is the mechanistic basis for each of the reactions you proposed in Exercise 7 that allows differentiation of 1° from 2° alkyl halides?

9. The LC_{50} (ppm) for inhalation of H-Br by rats is ________ over a time period of ________.

10. What toxic fumes are evolved in fires involving methyltrioctylammonium chloride and 1-bromohexane?

NAME (print): ______________________________ DATE: __________

INSTRUCTOR: ______________________________ LABORATORY SECTION: __________

1. Why should apparatus containing residues of bromine *not* be rinsed with acetone? How can such residues be chemically removed?

2. What should you do if any bromine comes in contact with your skin?

3. What is the vapor pressure of bromine at 20 °C?

4. Why should inhalation of the vapors of bromine be avoided?

5. What changes, if any, in the color of the reaction mixture do you anticipate as the bromination proceeds?

6. Underline the proper category of the bromination reaction: nucleophilic substitution, electrophilic substitution, nucleophilic addition, electrophilic addition, elimination..

7. Classify the overall reaction as to whether it is an oxidation, reduction, or neither. Show how you reached your conclusion.

8. The flash point (°C) of dichloromethane is ________.

9. The oral LD_{50} (mg/kg) in rats of bromine is ________.

NAME (print): ______________________________ DATE: __________

INSTRUCTOR: ______________________________ LABORATORY SECTION: __________

1. Why is it important to add the concentrated sulfuric acid *to* water rather than the reverse?

2. What is the function of sulfuric acid in promoting the hydration of norbornene?

3. Why would concentrated hydrochloric acid be an unsuitable replacement for sulfuric acid?

4. How much potassium hydroxide is required to neutralize 2 mL of *concentrated* sulfuric acid? Show your calculation.

5. Write a balanced equation for the chemical reaction(s) responsible for the pressure build-up when the crude reaction mixture is washed with aqueous sodium bicarbonate.

6. What is the composition of the solid that might appear in the separatory funnel during the work-up?

7. What is the purpose of washing the ethereal layer containing the product with saturated sodium bicarbonate and sodium chloride prior to drying the solution with anhydrous sodium sulfate?

8. Why is it necessary to remove all of the diethyl ether prior to subliming the product?

9. What is the purpose of the trap between the water aspirator and the sublimation apparatus?

10. Why is it necessary to seal the capillary tube before determining the melting point of *exo*-norborneol?

11. In rats, the oral LD_{50} (mg/kg) of sulfuric acid is ________.

12. The flash point (°C) of norbornene is ________ and its oral LD_{50} (mg/kg) in rats is ________.

Pre-Lab Exercises for HYDROBORATION-OXIDATION OF (+)-α-PINENE (Section 10.7)

NAME (print): ______________________________ DATE: ____________

INSTRUCTOR: ______________________________ LABORATORY SECTION: __________

1. Determine the molar ratios of each of the reactants to be used in this procedure and specify the "limiting reagent." Show your calculations.

2. Write the chemical equation by which THF disrupts the usual equilibrium between diborane and borane.

3. Why are the solvent and apparatus to be carefully dried prior to the reaction?

4. A the end of the reaction, water is to be added to the reaction mixture containing the borane-THF complex. Why is this addition to be *slow*?

5. Write a balanced equation for the reaction that occurs between water and borane-THF complex.

6. What purpose is served by adding basic hydrogen peroxide to the reaction mixture?

7. What should you do if hydrogen peroxide comes in contact with your skin?

8. Write equations for any chemical test(s) you might use to determine whether the desired alcohol is contaminated with (+)-α-pinene.

9. Underline the media appropriate for extinguishing fires involving borane/THF solutions: Water Carbon dioxide Chemical powder Foam

10. In rats, the oral LD_{50} (mg/kg) of tetrahydrofuran is ________.

NAME (print): ______________________ DATE: __________

INSTRUCTOR: ______________________ LABORATORY SECTION: __________

1. Explain the purpose of placing a carborundum boiling stone in the reaction vessel prior to heating.

2. Explain why triethylene glycol is used as the reaction solvent for this reaction.

3. Calculate the molar ratio of base to the *meso*-stilbene dibromide used in this experiment. For the purposes of this calculation, assume that the KOH contains 15% water.

4. Explain why a sand bath is preferred to a mineral oil bath as the heating source for this experiment.

5. In this experiment, the reaction vessel is cooled to room temperature prior to adding water. Explain why this is done rather than adding water to the hot reaction mixture.

6. Diphenylacetylene is only sparingly soluble in diethyl ether. Explain why is the addition of water to the reaction mixture is preferable to adding diethyl ether as a means of precipitating the product for isolation by filtration.

7. The flash point (°C) of triethylene glycol is________ and its oral LD_{50} (mg/kg) in rats is ________.

Pre-Lab Exercises for PREPARATION OF 3-METHYL-3-HYDROXY-2-BUTANONE (Section 11.3)

NAME (print): ______________________________ DATE: __________

INSTRUCTOR: ______________________________ LABORATORY SECTION: __________

1. Write the structural formula for the product of the reaction of 2-methyl-3-butyn-2-ol with bromine in carbon tetrachloride.

2. In the preparation of the reagents for hydration of 2-methyl-3-butyn-2-ol, why is concentrated sulfuric acid added to water rather than water added to sulfuric acid?

3. What is the purpose of adding potassium carbonate *and* sodium chloride to the distillate before extracting it with dichloromethane? Why not use just potassium carbonate?

4. Calculate the molar ratio of mercuric oxide and 2-methyl-3-butyn-2-ol used in this experiment. Show your calculations.

5. Is mercuric oxide the limiting reagent? Explain.

6. Why does hydration of 2-methyl-3-butyn-2-ol give the ketone, 3-hydroxy-3-methyl-2-butanone (**4**), rather than the aldehyde, 3-methyl-3-hydroxybutanal (**5**)?

7. How can **4** and **5** be differentiated chemically? by spectroscopic methods?

8. Underline the media, if any, that are *not* appropriate for extinguishing fires involving 2-methyl-3-butyn-2-ol: Water Carbon dioxide Chemical powder Foam

9. The flash point (°C) of 2-methyl-3-butyn-2-ol is ________.

Pre-Lab Exercises for FORMATION OF A SILVER ACETYLIDE AND ITS DECOMPOSITION (Section 11.4)

NAME (print): ______________________________ DATE: __________

INSTRUCTOR: ______________________ LABORATORY SECTION: __________

1. Write a balanced equation for the reaction of silver ammonia complex with 2-methyl-3-butyn-2- ol.

2. Why is the silver salt of the alkyne not allowed to dry on the filter?

3. Write an equation for the reaction of dilute hydrochloric acid with the silver salt of the alkyne.

4. Put a check mark beside the reported chronic effects, if any, of silver nitrate: carcinogen ________, mutagen ________, tumorigen ________.

5. The LHL_O (ppm) for inhalation of ammonium hydroxide by humans is ________ over a time period of________; the oral LD_{50} (mg/kg) in rats of 2-methyl-3-butyn-2-ol is ________.

Pre-Lab Exercises for DIELS-ALDER REACTION. Part A. REACTION OF 1,3-BUTADIENE AND MALEIC ANHYDRIDE (Section 12.3)

NAME (print): ______________________ DATE: __________

INSTRUCTOR: ______________________ LABORATORY SECTION: __________

1. Specify the conformation required to enable a 1,3-diene to undergo the Diels-Alder reaction?

2. Which conformation of 1,3-butadiene, *s-cis* or *s-trans,* is thermodynamically preferred and why?

3. Why does 1,3-butadiene react more rapidly with maleic anhydride than with another molecule of itself?

4. Why is 3-sulfolene rather than 1,3-butadiene itself to be used as a source of the diene?

5. Define the term "*in situ* preparation" as it applies to this experiment.

6. Why is a gas trap required in this experiment?

7. What is the limiting reagent in the reaction between 1,3-butadiene and maleic anhydride and why do you think this reagent rather than the other was made limiting? Show your calculations.

8. Why is petroleum ether to be added to the solution of product in xylene?

9. What type of filtration is to be used to isolate the crystalline product?

10. The oral LD_{50} (mg/kg) in rats of 3-sulfolene is ________; that of maleic anhydride is________.

11. The flash point (°C) of xylene is________.

12. List the target organs for xylene.

13. Underline the media, if any, that are *not* appropriate for extinguishing fires involving xylene:

Water Carbon dioxide Chemical powder Foam

Pre-Lab Exercises for DIELS-ALDER REACTION. Part B. REACTION OF 1,3-CYCLO-PENTADIENE AND MALEIC ANHYDRIDE (Section 12.3)

NAME (print): ______________________________ DATE: __________

INSTRUCTOR: ______________________ LABORATORY SECTION: ________

1. What conformation is required to enable a 1,3-diene to undergo the Diels-Alder reaction?

2. What is the limiting reagent in the reaction between 1,3-cyclopentadiene and maleic anhydride, and why do you think this reagent rather than the other one was made limiting? Show your calculations.

3. Why does 1,3-cyclopentadiene react more rapidly with maleic anhydride than with another molecule of itself?

4. Why is a 25-mL rather than a 10-mL round-bottom flask to be used for cracking dicyclopentadiene when only 7 mL of this substance is being cracked?

5. Why is a fractional rather than a simple distillation apparatus specified for the cracking of dicyclopentadiene?

6. Why is the solution of maleic anhydride in ethyl acetate-petroleum ether to be homogeneous when 1,3-cyclopentadiene is added to it?

7. Why is the solution containing the desired product to be cooled slowly?

8. What type of filtration, gravity or vacuum, is to be used for isolating the product?

9. The oral LD_{50} (mg/kg) in rats of dicyclopentadiene is ________ whereas that of maleic anhydride is ________.

10. Maleic anhydride is listed as a potentially carcinogenic compound. T ______ F ______.

11. The flash points (°C) of dicyclopentadiene and ethyl acetate are________ and________, respectively.

12. List the acute effects of exposure to vapors or mists containing dicyclopentadiene.

Pre-Lab Exercises for DIELS-ALDER REACTION. Part B. REACTION OF 1,3-CYCLOPENTADIENE AND MALEIC ANHYDRIDE (Section 12.3)

NAME (print): ______________________________ DATE: __________

INSTRUCTOR: ______________________________ LABORATORY SECTION: __________

1. What conformation is required to enable a 1,3-diene to undergo the Diels-Alder reaction?

2. What is the limiting reagent in the reaction between 1,3-cyclopentadiene and maleic anhydride, and why do you think this reagent rather than the other one was made limiting? Show your calculations.

3. Why does 1,3-cyclopentadiene react more rapidly with maleic anhydride than with another molecule of itself?

4. Why is the solution of maleic anhydride in ethyl acetate-petroleum ether to be homogeneous when 1,3-cyclopentadiene is added to it?

5. Why is the solution containing the desired product to be cooled slowly?

6. What technique is to be used to effect slow cooling of the solution of product?

7. What type of filtration, gravity or Craig tube, is to be used for isolating the product?

8. Write equations for the chemical tests you are to perform on the product.

9. The oral LD_{50} (mg/kg) in rats of ethyl acetate is ________ whereas that of maleic anhydride is ________.

10. Maleic anhydride is listed as a potentially carcinogenic compound. T ______ F ______.

11. The flash point (°C) of ethyl acetate is ________.

NAME (print): ______________________________ DATE: ____________

INSTRUCTOR: ______________________________ LABORATORY SECTION: ____________

1. What serves as the nucleophile in the conversion of a cyclic anhydride to a dicarboxylic acid?

2. Cyclic anhydrides are often insoluble in water, whereas the dicarboxylic acids derived from them are soluble. Why is there a difference in the water-solubilities of these two classes of compounds?

3. Why should a warm aqueous solution of a solid *not* be cooled in an ice-water bath before crystallization has started?

4. What strategies might you use to induce crystallization if the desired product oils out rather than crystallizes?

5. What do you expect to observe when an aqueous solution of your product is tested with pHydrion paper?

6. What do you expect to observe when you treat your product with Br_2/CH_2Cl_2?

7. Write the structure of the product expected from the reaction between your product and Br_2/CH_2Cl_2.

Pre-Lab Exercises for FORMATION OF SEMICARBAZONES UNDER KINETIC AND THERMODYNAMIC CONTROL (Section 13.2)

NAME (print): ______________________________ DATE: ____________

INSTRUCTOR: ______________________________ LABORATORY SECTION: __________

1. Define:

a. kinetic control of a reaction.

b. thermodynamic control of a reaction.

2. Write a balanced equation for the reaction between semicarbazide hydrochloride and dibasic potassium phosphate.

3. The three buffer systems described in this chapter are **A**: $CH_3CO_2H/CH_3CO_2^-$; **B**: $H_2PO_4^-/HPO_4^{2-}$; and **C**: H_2CO_3/HCO_3^-. Answer the following questions by using A, B, or C to refer to the proper buffer system.

a. Which buffer system maintains the lowest pH?______

b. Which buffer system generally produces the fastest rate of formation of semicarbazones of aldehydes and ketones? ______

c. Which buffer system maintains a pH in the range 6.1–6.2? ______

4. What is the purpose of cooling each reaction mixture in an ice-water bath after reaction periods of various times and at various temperatures?

5. Typical melting ranges of crystals obtained in various parts of this experiment might be, for example, 150–162 °C, 163–165 °C, and 199–201 °C. Identify these melting ranges with the mixtures of semicarbazones of cyclohexanone (C) and 2-furaldehyde (F) listed below.

   a. 98% C/2%F ______________________

   b. 98% F/2% C ______________________

   c. 40% F/60% C ______________________

6. Why should a Thiele tube containing miner oil *not* be used for determining melting points in this experiment?

7. The flash points (°C) of cyclohexanone and 2-furaldehyde are________ and________, respectively.

8. The LC_{50} (ppm) for inhalation by rats of cyclohexanone is ________ during________ hours, whereas that of 2-furaldehyde is ________ during________ hours.

9. List the toxic fumes evolved in fires involving the semicarbazones of cyclohexanone and 2-furaldehyde.

10. Specify whether 2-furaldehyde is listed as tumorigenic compound.

11. List the target organs for cyclohexanone.

12. List the target organs for 2-furaldehyde.

NAME (print): ______________________________ DATE: __________

INSTRUCTOR: ______________________________ LABORATORY SECTION: __________

1. Describe the method by which hydrogen bromide is prepared for use in this experiment.

2. Determine the limiting reagent for the conversion of 1-butanol to 1-bromobutane and the theoretical yield of product.

3. Is this reaction an S_N1 or S_N2 process? How will the mechanism be confirmed experimentally?

4. Write the structures of two possible side products that might be formed in this reaction.

5. What is the purpose of the following experimental techniques, and where is each used in the preparation?

a. heating at reflux

b. simple distillation

c. adding anhydrous sodium sulfate

6. The oral LD_{50} (mg/kg) in rats of 1-butanol is ________; the intraperitoneal LD_{50} (mg/kg) in rats of 1-bromobutane is ________.

7. The flash points (°C) of the compounds listed in Exercise 6 are ________ and________, respectively.

8. What toxic fumes are evolved in fires involving 1-bromobutane?

9. Indicate whether the following statement is true (T) or false (F): Water is an appropriate medium for extinguishing fires involving 1-bromobutane.

____

NAME (print): ______________________________ DATE: ____________

INSTRUCTOR: ______________________________ LABORATORY SECTION: __________

1. Determine the limiting reagent for the preparation of 2-chloro-2-methylpropane and calculate the theoretical yield of product.

2. Why is this reaction carried out at room temperature rather than at elevated temperatures?

3. Is this reaction an S_N1 or S_N2 process? Does the experiment contain any procedures for experimental verification of the mechanism?

4. Write the structures of any possible side products in the reaction.

5. After the initial reaction is carried out and the crude product is isolated, it is washed with saturated sodium bicarbonate solution. What is the purpose of this wash, and could dilute sodium hydroxide solution be used instead of sodium bicarbonate? Explain briefly.

6. The flash point (°C) of 2-methyl-2-butanol is ________.

7. List the toxic fumes evolved in fires involving 2-chloro-2-methylbutane.

8. List the media appropriate for extinguishing fires involving 2-methyl-2-butanol.

Pre-Lab Exercises for FRIEDEL-CRAFTS ALKYLATION OF *p*-XYLENE WITH 1-BROMOPROPANE (Section 15.2)

NAME (print): ______________________________ DATE: __________

INSTRUCTOR: ______________________________ LABORATORY SECTION: __________

1. **a.** Determine the limiting reagent in this experiment. Show your calculations.

   b. Why is this reagent chosen to be limiting?

2. **a.** What function is served by $AlCl_3$ in this reaction?

   b. Why is so little of it required?

3. Why should the *p*-xylene and the apparatus to be used in the experimental procedure be dry?

4. What experimental technique is to be used for drying the *p*-xylene?

5. Do you expect the ratio of *n*-propyl-*p*-xylene to isopropyl-*p*-xylene produced in this experiment to be larger or smaller than the ratio of *n*-propylbenzene to isopropylbenzene produced by a similar alkylation of benzene by 1-bromopropane? Explain your answer.

6. Why is the 1-bromopropane to be added dropwise to the reaction mixture rather than all at once?

7. What is the reason for pouring the completed reaction mixture into a crushed ice/water mixture? Why water, and why ice?

8. What compounds are present in the aqueous solution that is to be discarded?

9. What hazardous combustion products are evolved in fires involving $AlCl_3$, *p*-xylene, and 1-bromopropane?

10. The flash points (°C) of *p*-xylene and 1-bromopropane are ________ and ________, respectively.

11. List the media appropriate for extinguishing fires involving *p*-xylene and 1-bromopropane.

Pre-Lab Exercises for ACYLATION OF *m*-XYLENE WITH PHTHALIC ANHYDRIDE (Section 15.3)

NAME (print): ______________________________ DATE: ____________

INSTRUCTOR: ______________________________ LABORATORY SECTION: __________

1. Write a stepwise mechanism for the reaction between *m*-xylene (**13**) and the acylium ion **16** to the salt **14**; specify the rate-determining step in your mechanism. Use curved arrows to symbolize the flow of electrons.

2. Why does ion **16** attack at C-4 rather than C-2 of *m*-xylene?

3. Why is a gas trap a necessary part of the apparatus for this procedure?

4. Why may an ice-water bath be needed after the reaction begins?

5. Why should solvent-grade rather than anhydrous diethyl ether be used for extracting the solution obtained by hydrolysis of the reaction mixture?

6. List the target organs for toxic effects of *m*-xylene.

7. What toxic fumes are evolved in fires involving $AlCl_3$?

8. The values of the oral LD_{50} (mg/kg) in rats of $AlCl_3$, *m*-xylene, and phthalic anhydride are ________, ________, and ________, respectively.

NAME (print): ______________________________ DATE: __________

INSTRUCTOR: ______________________________ LABORATORY SECTION: __________

1. Calculate the molar ratio of nitric acid:bromobenzene used in this experiment. (Concentrated nitric acid is 16 *M*.) What is the limiting reagent in this experiment?

2. Why is dinitration not a significant process under the conditions to be used?

3. Compare the ratio of reactants in this experiment (Exercise 1, above) with that in the experiment of Section 15.2 (see Pre-Lab Exercise 1 for that section). Why are they so different?

4. What is the function of the concentrated sulfuric acid in this experiment?

5. Why is the reaction mixture to be stirred during the addition of bromobenzene to the mixture of acids?

6. Is the nitration reaction expected to be exothermic or endothermic?

7. What compounds are present in the aqueous solution from which the isomeric bromonitrobenzenes are to be filtered?

8. Why is the *p*-isomer of the product expected to be less soluble in ethanol than the *o*-isomer?

9. Underline the media, if any, that are *not* appropriate for extinguishing fires involving bromobenzene and *o*- and *p*-bromonitrobenzene: Water Carbon dioxide Chemical powder Foam

10. List the toxic fumes evolved in fires involving bromobenzene and *o*- and *p*-bromonitrobenzene.

Pre-Lab Exercises for NITRATION OF BROMOBENZENE. Part B. THIN-LAYER CHROMATOGRAPHY (Section 15.4)

NAME (print): ______________________ DATE: __________

INSTRUCTOR: ______________________ LABORATORY SECTION: __________

1. Why should the developing chamber for a TLC plate *not* be open to the atmosphere?

2. Why should a TLC plate be removed from the solvent before the solvent front reaches the top of the plate?

3. Which of the following diagrams illustrate(s) an *improper* way of spotting a TLC plate? Tell what is wrong in each such case.

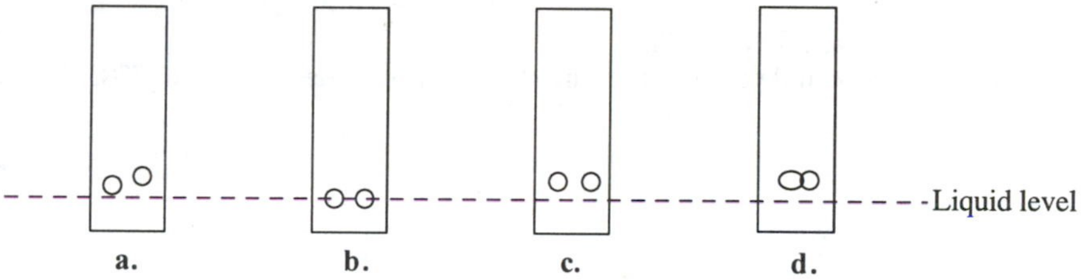

a.

b.

c.

d.

5. What problem would attend failure to mark the position of the solvent front on the TLC plate immediately after developing the plate?

6. What technique are you to use to determine the location of compounds on the TLC plate once it has been developed?

7. What consequence would you predict if a more polar eluant were used for the TLC chromatography?

8. The flash points (°C) for ethyl acetate and hexane are________ and________, respectively.

Pre-Lab Exercises for NITRATION OF BROMOBENZENE. Part C. COLUMN CHROMATOGRAPHY (Section 15.4)

NAME (print): ____________________ DATE: __________

INSTRUCTOR: ____________________ LABORATORY SECTION: __________

1. What difficulty may result if

   a. the chromatography column is not placed in a vertical position?

   b. the liquid level of the eluent is allowed to drop below the top of the packing in the column?

2. Why should there be no air bubbles in the column after it has been packed?

3. What is the purpose of the layer of sand added at the top of the adsorbent when preparing the chromatography column?

4. Why should a minimum amount of solvent be used to introduce a mixture onto the chromatography column?

5. What consequence would you predict if a more polar eluant were used for the column chromatography?

6. What technique are you to use to monitor the contents of the various fractions of eluant that you collect?

7. The flash points (°C) for ethyl acetate and hexane are________ and________, respectively.

Pre-Lab Exercises for RELATIVE RATES OF ELECTROPHILIC AROMATIC SUBSTITUTION (Section 15.5)

NAME (print): ______________________________ DATE: __________

INSTRUCTOR: ______________________________ LABORATORY SECTION: __________

1. Why is 15 *M* acetic acid an appropriate solvent in which to perform rate studies of electrophilic brominations?

2. Show that 15 *M* acetic acid is approximately 90% acetic acid and 10% water by weight.

3. Why is benzene itself not to be used as a substrate in this experiment?

4. Why is it important to add the solution of bromine quickly and in one portion to the test tube containing the aromatic substrate?

5. What criterion is to be used for measuring the rates of bromination in the experiment?

6. Why is it important to maintain a constant temperature in this experiment?

7. Why might it be necessary to conduct this experiment at more than one temperature?

8. The flash points (°C) of anisole, diphenyl ether, and phenol are ________, ________, and ________, respectively.

9. Underline the media, if any, that are *not* appropriate for extinguishing fires involving the compounds listed in Exercise 8: Water Carbon dioxide Chemical powder Foam

Pre-Lab Exercises for OXIDATION OF ALCOHOLS. Part A. OXIDATION OF CYCLODODECANOL TO CYCLODODECANONE (Section 16.2)

NAME (print): ______________________________ DATE: ____________

INSTRUCTOR: ______________________________ LABORATORY SECTION: __________

1. What is the oxidizing agent in this experiment and to what is it reduced during the course of the reaction?

2. What qualitative test is performed to determine whether sufficient oxidizing agent has been used in this procedure?

3. Write an equation that expresses the chemical reaction that forms the basis of the test of Exercise 2.

4. What purpose is served by washing the ethereal extract containing the product with saturated sodium bicarbonate?

5. What purpose is served by washing the ethereal extract containing the product with saturated sodium bisulfite?

6. The flash points (°C) of acetone and acetic acid are ________, and ________, respectively.

7. An MSDS is not presently available on either cyclododecanol or cyclododecanone. In lieu of this information, provide the relevant toxicological information for the analogous compounds, cyclohexanol and cyclohexanone.

8. Underline the media, if any, that are *not* appropriate for extinguishing fires involving cyclododecanol and cyclododecanone (assume these are the same as for cyclohexanol and cyclohexanone, respectively):
Water Carbon dioxide Chemical powder Foam

Pre-Lab Exercises for OXIDATION OF ALCOHOLS. Part B. OXIDATION OF 4-CHLOROBENZYL ALCOHOL TO 4-CHLOROBENZOIC ACID (Section 16.2)

NAME (print): ______________________________ DATE: __________

INSTRUCTOR: ______________________________ LABORATORY SECTION: __________

1. What is the oxidizing agent in this experiment and to what is it reduced in the course of the reaction?

2. What qualitative test is performed to determine whether sufficient oxidizing agent has been used in this procedure?

3. Write an equation that expresses the chemical reaction that forms the basis of the test of Exercise 2.

4. What purpose is served by washing the ethereal extract containing the product with saturated sodium bicarbonate?

5. The flash points (°C) of acetonitrile, acetic acid, and diethyl ether are ________, and ________, and ________, respectively.

6. Thermal decomposition of calcium hypochlorite occurs at approximately ________°C.

7. The oral LD_{50} (mg/kg) in rats of 4-chlorobenzoic acid is ________; of calcium hypochlorite is ________.

8. Underline the media, if any, that are *not* appropriate for extinguishing fires involving 4-chlorobenzyl alcohol and 4-chlorobenzoic acid.

Water   Carbon dioxide   Chemical powder   Foam

Pre-Lab Exercises for BASE-CATALYZED OXIDATION-REDUCTION OF ALDEHYDES BY THE CANNIZZARO REACTION (Section 16.3)

NAME (print): ______________________________ DATE: __________

INSTRUCTOR: ______________________________ LABORATORY SECTION: __________

1. Write the balanced equation for the oxidation-reduction of 4-chlorobenzaldehyde to 4-chlorobenzyl alcohol and potassium 4-chlorobenzoate by reaction with potassium hydroxide. Show all work.

2. Why is the Cannizzaro reaction limited to aldehydes having no α-hydrogen atoms?

3. What is the limiting reagent in this reaction? Show how you arrived at this conclusion.

4. Why do ketones having no α-hydrogen atoms not undergo the Cannizzaro reaction?

5. What is an emulsion? Why is it desirable to have an emulsion rather than two phases during the reaction between 4-chlorobenzaldehyde and aqueous base?

6. What is the solid that is formed after 4-chlorobenzaldehyde has been allowed to react with aqueous potassium hydroxide? Why does it dissolve in water?

7. List the hazardous combustion or decomposition products of 4-chlorobenzaldehyde.

8. The oral LD_{50} (mg/kg) in rats of 4-chlorobenzaldehyde is________ and that of 4-chlorobenzoic acid is ________.

Pre-Lab Exercises for HYDROGENATION OF 4-CYCLOHEXENE-*cis*-1,2-DICARBOXYLIC ACID (Section 17.2)

NAME (print): ______________________ DATE: __________

INSTRUCTOR: ______________________ LABORATORY SECTION: __________

1. What is meant by the term *catalytic hydrogenation*?

2. What catalyst is used in this preparation, and how is it prepared?

3. Indicate how hydrogen gas is generated for the preparation of the catalyst; write a balanced equation to show this reaction.

4. What is the stereochemistry of catalytic hydrogenation? Cite an example that shows this.

5. Does the catalytic hydrogenation of 4-cyclohexene-*cis*-1,2-dicarboxylic acid show the stereochemistry of hydrogenation? Explain.

6. Aqueous acid is used as the solvent for the reaction. Could water at pH = 7 have been used as the solvent instead of the one that is used? Explain.

7. Does addition of concentrated HCl to an aqueous solution of the reduction product of this reaction increase or decrease the water solubility of the product? Why?

8. What hazardous mineral acid is produced by decomposition of chloroplatinic acid?

9. Underline the media appropriate for extinguishing fires involving sodium borohydride:

Water Carbon dioxide Chemical powder Foam

10. What MSDS toxicological data, if any, are available for 4-cyclohexene-1,2-dicarboxylic acid?

Pre-Lab Exercises for FORMATION AND REDUCTION OF *N*-CINNAMYLIDENE-*m*-NITROANILINE (Section 17.3)

NAME (print): ______________________________ DATE: __________

INSTRUCTOR: ______________________________ LABORATORY SECTION: __________

1. Draw the structure of an imine, and give the equation for its conversion into an amine as done in this experiment.

2. What metal hydride reducing agent is used in this experiment, and what are the advantages of using it rather than hydrogen gas in the presence of a catalyst? What product would be formed if catalytic hydrogenation were used?

3. Define the term *reductive amination.*

4. Draw the structure of the imine intermediate that is formed, and describe what experimental technique is used to maximize the yield of this compound.

5. Why is it unnecessary to isolate and purify the intermediate imine?

6. Determine the limiting reagent in this preparation and compute the theoretical yield. Show all work.

7. List the toxic fumes evolved in a fire involving cinnamaldehyde, *m*-nitroaniline, and cyclohexane.

8. The oral LD_{50} (mg/kg) in rats of *m*-nitroaniline is ________.

9. The flash points (°C) of cyclohexane and cinnamaldehyde are ________ and ________, respectively.

10. Underline the media appropriate for extinguishing fires involving sodium borohydride:

Water Carbon dioxide Chemical powder Foam

11. Write an equation for reaction of sodium borohydride and methanol that yields hydrogen.

NAME (print): ______________________________ DATE: __________

INSTRUCTOR: ______________________________ LABORATORY SECTION: __________

1. Determine whether sodium borohydride is the limiting reagent in this experiment. Show your work.

2. Why is it important to avoid exposing sodium borohydride to moisture? Write the reaction that occurs when sodium borohydride and water are mixed.

3. What color change should occur as the reduction of fluorenone proceeds? Provide a reason for this change.

4. In the experiment, the reducing agent is allowed to react with fluorenone, and after the reaction is complete, sulfuric acid is added. What is the purpose of this addition? Why is it important to dissolve all of the solids completely?

5. Describe the advantages of using a metal hydride reduction reaction rather than catalytic hydrogenation in this procedure.

6. Write an equation for the reaction that might occur during the recrystallization of fluorenol from methanol if all of the sulfuric acid has not been removed by washing.

7. What toxicological data are available on the MSDS for fluorenol? for fluorenone?

8. Underline the media, if any, that are *not* appropriate for extinguishing fires involving a combination of fluorenone, fluorenol, methanol, and sodium borohydride:

Water   Carbon dioxide   Chemical powder   Foam

NAME (print): ______________________________ DATE: __________

INSTRUCTOR: ______________________________ LABORATORY SECTION: __________

1. **Why should the Erlenmeyer flask in which the fermentation is conducted not be stoppered tightly?**

2. **What is the role of the Na_2HPO_4?**

3. **What is the practical advantage of using dichloromethane as the extraction solvent rather than diethyl ether?**

4. **Why use filter-aid for filtering the reaction mixture?**

5. **Why is it important *not* to shake the separatory funnel vigorously during the extraction of the product with dichloromethane?**

6. Although sucrose is added to the reaction, it is not necessary. Propose a reason for using sucrose, and account for the fact that it is not absolutely required for the reduction.

7. The oral LD_{50} (mg/kg) in rats of sucrose is ________, that of methyl acetoacetate is ________.

8. What toxic fumes are evolved upon combustion of dichloromethane?

Pre-Lab Exercises for DETERMINING OPTICAL PURITY OF METHYL (*S*)-3-HYDROXYBUTANOATE (Section 17.6)

NAME (print): ______________________ DATE: ____________

INSTRUCTOR: ______________________ LABORATORY SECTION: __________

1. Why is it important to use a dry NMR tube for measuring the optical purity of a sample using a chiral shift reagent?

2. Why should the chiral shift reagent be stored in a desiccator?

3. What is the effect of having undissolved solid in the NMR sample?

4. Why should the sample be allowed to stand for about 20 min after dissolution of the shift reagent?

5. What is the estimated accuracy of determining the enantiomeric excess by the NMR method?

6. When determining the optical purity of an alcohol using chiral shift reagents, it is important to perform the experiment for samples of both the racemic alcohol and enantiomerically enriched alcohol. Explain.

7. The ratio of the two enantiomers is determined by comparing the peak areas for the methyl peaks on the ester group. Why is this better than trying to compare the peak areas for the methine hydrogen a to the hydroxyl group?

8. List the toxic products evolved in fires involving chloroform.

9. Indicate whether the following statement is true (T) or false (F): Chloroform is listed as a potential carcinogen. ________

Pre-Lab Exercises for WITTIG AND WADSWORTH-EMMONS REACTIONS. Part A.
SYNTHESIS OF (*Z*)- AND (*E*)-STILBENES BY A WITTIG REACTION.
(Section 18.2)

NAME (print): ______________________________ DATE: __________

INSTRUCTOR: ______________________________ LABORATORY SECTION: __________

1. What is the limiting reagent in this experiment? Show your work.

2. Explain why it is important to stir the reaction mixture vigorously.

3. After the Wittig reaction is complete, the solution of isomeric stilbenes in dichloromethane is washed with saturated sodium bisulfite. What is the purpose of this wash?

4. Suppose the aldehyde used in a Wittig reaction is contaminated with the corresponding carboxylic acid. What complication would this cause?

5. After the irradiation of the isomeric stilbenes in dichloromethane containing iodine is complete, the solution is washed with saturated sodium bisulfite. What is the purpose of this wash?

6. The oral LD_{50} (mg/kg) in rats of benzaldehyde is ________; the oral LD_{50} (mg/kg) of (*E*)-stilbene in mice is ________; of triphenylphosphine oxide is ________.

7. List the organs for which benzyltriphenylphosphonium chloride is irritating.

8. List the target organs of dichloromethane.

Pre-Lab Exercises for WITTIG AND WADSWORTH-EMMONS REACTIONS. Part B. SYNTHESIS OF (*E*)-STILBENE BY THE WADSWORTH-EMMONS REACTION. (Section 18.2)

NAME (print): ______________________________ DATE: __________

INSTRUCTOR: ________________________ LABORATORY SECTION: ________

1. What is the limiting reagent in this experiment? Show your work.

2. Explain why exposure of the solution of potassium *tert*-butoxide in *N,N,*-dimethylformamide to the atmosphere should be minimized.

3. Suppose the aldehyde used in a Wadsworth-Emmons reaction is contaminated with the corresponding carboxylic acid. What complication would this cause?

4. Is the reaction of the potassium salt **13** of the phosphonate ester with benzaldehyde exothermic or endothermic? Give the basis for your answer.

5. How is the product of this experiment freed from residual benzaldehyde?

6. The oral LD_{50} (mg/kg) in rats of benzaldehyde is ________; the oral LD_{50} (mg/kg) of (*E*)-stilbene in mice is ________; the intravenous LD_{50} (mg/kg) of diethyl benzylphosphonate in mice is ________.

7. Indicate whether the following statement is true (T) or false (F): *N,N*-Dimethylformamide is listed as a potential carcinogen. ________

8. The flash point (°C) of *N,N*-dimethylformamide is ________.

Pre-Lab Exercises for SYNTHESIS OF *TRANS-p*-ANISAL-ACETOPHENONE (Section 18.3)

NAME (print): ______________________________ DATE: ____________

INSTRUCTOR: ______________________________ LABORATORY SECTION: __________

1. *p*-Anisaldehyde and acetophenone are used in equimolar amounts in this experiment. Suggest a complication that might result if two moles of ketone per mole of aldehyde were used.

2. The amount of sodium hydroxide used to promote the condensation reaction is less than an equimolar amount. What complication might result if a much larger amount of concentrated sodium hydroxide were used?

3. Why is the condensation (dehydrated) product rather than the aldol addition (hydrated) product obtained in this experiment?

4. Why does the enolate ion of an aromatic ketone react faster with an aldehyde group, producing a crossed-aldol reaction, than with the carbonyl group of another molecule of ketone?

5. The oral LD_{50} (mg/kg) in rats of acetophenone is ________; the corresponding value for *p*-anisaldehyde is________.

Pre-Lab Exercises for PREPARATION OF 4,4-DIMETHYL-2-CYCLOHEXEN-1-ONE (Section 18.4)

NAME (print): ______________________________ DATE: __________

INSTRUCTOR: ______________________________ LABORATORY SECTION: __________

1. Why is an *intramolecular* rather than an *inter*molecular aldol condensation favored with 2,2-dimethylhexanal-5-one (**28**)?

2. What is the limiting reagent in the reaction to form the Michael addition product **28**? Show your work.

3. Why do you think the carbonyl-containing compound identified in Exercise 2 rather than the other was made limiting?

4. Why must water be removed from the reaction mixture during the course of the reaction?

5. 1-Propanol boils at only a slightly lower temperature than toluene, and all the reagents used in this experiment are soluble in it. Why is it an inappropriate solvent to use in this procedure?

6. This reaction occurs much faster if a sulfonic acid rather than a carboxylic acid is used as the catalyst. Why might this be?

7. What is the theoretical volume of water that would be obtained in the reaction? Show your work. How might you account for the production of *greater* than a theoretical volume in the experiment?

8. The flash points (°C) of toluene, 3-buten-2-one, and 2-methylpropanal are ________, ________, and ., respectively.

9. The oral LD_{50} (mg/kg) in rats of 3-buten-2-one is ________; the corresponding value for 2-methyl-propanal is ________.

10. Indicate whether the following statement is true (T) or false (F): 3-Buten-2-one is listed as a potential carcinogen. ________

Pre-Lab Exercises for PREPARATION AND REACTIONS OF GRIGNARD REAGENTS.
Part A. PREPARATION OF THE GRIGNARD REAGENT
(Section 19.2)

NAME (print): ______________________________ DATE: __________

INSTRUCTOR: ______________________________ LABORATORY SECTION: __________

1. What is the limiting reagent in this reaction? Show your work.

2. Why are ethereal solvents important to the success of preparing the Grignard reagent?

3. Why must the reagents, solvents, and apparatus used for preparing the Grignard reagent be dry?

4. Why were you cautioned not to place plastic and rubber parts in the drying oven?

5. Why is it necessary to have an ice-water bath available during the preparation of the Grignard reagent?

6. What signs should you look for in determining whether the reaction initiated?

7. Why should the aryl or alkyl halide from which the Grignard reagent is to be made *not* be added all at once to the reaction flask?

8. Why is anhydrous diethyl ether added to the magnesium in two portions, one at the beginning of the reaction and the second after formation of the Grignard reagent has started?

9. The flash points (°C) of diethyl ether, bromobenzene, and 1-bromobutane are ________, ________, and ________, respectively.

10. 1,2-Dibromoethane is *not* listed as a potential (underline as appropriate):

Carcinogen Hallucinogen Tumorigen

11. List the toxic fumes evolved in fires involving diethyl ether and either bromobenzene or 1-bromobutane.

Pre-Lab Exercises for PREPARATION AND REACTIONS OF GRIGNARD REAGENTS.
Part B. PREPARATION OF TRIPHENYLMETHANOL
(Section 19.2)

NAME (print): ______________________ DATE: __________

INSTRUCTOR: ______________________ LABORATORY SECTION: __________

1. What is the limiting reagent in this reaction? Show your work.

2. Why do you think the particular reagent specified in Exercise 1 was made limiting?

3. Why would the presence of methanol in the methyl benzoate lower the yield of triphenylmethanol?

4. Why must the ester and the diethyl ether in which it is dissolved be anhydrous?

5. Why is the solution of ester added to the Grignard reagent in a dropwise fashion rather than all at once?

6. What is wrong with storing ethereal solutions in the laboratory bench from one period to the next?

7. Why is technical rather than anhydrous diethyl ether used in the work-up procedure?

8. Why is saturated aqueous sodium bicarbonate rather than just water used to remove residual sulfuric acid from the organic solution during the work-up procedure?

9. How are unchanged ester and biphenyl removed from the desired product by the work-up procedure used?

10. What happens to any Grignard reagent that remains in the reaction mixture after addition of the ester?

11. The flash point (°C) of methyl benzoate is ________; its oral LD_{50} (mg/kg) in rats is ________.

12. The flash point (°C) of cyclohexane is________.

Pre-Lab Exercises for PREPARATION AND REACTIONS OF GRIGNARD REAGENTS.
Part C. PREPARATION OF BENZOIC ACID (Section 19.2)

NAME (print): ______________________________ DATE: ____________

INSTRUCTOR: ______________________________ LABORATORY SECTION: __________

1. What is the limiting reagent in this procedure? Show your work.

2. Why do you think the particular reagent specified in Exercise 1 was made limiting?

3. What would the effect on yield be of allowing the crushed Dry Ice to be exposed to the atmosphere for long periods of time before use?

4. Why does pressure develop in the separatory funnel when the ethereal solution of benzoic acid is being extracted with aqueous sodium hydroxide?

5. Why is technical rather than anhydrous diethyl ether used in the work-up of the reaction mixture?

6. How are possible by-products such as biphenyl and benzophenone removed from the desired benzoic acid by the work-up procedure?

7. What is wrong with storing ethereal solutions in the laboratory bench from one period to the next?

8. The oral LDL_0 (mg/kg) in humans of benzoic acid is ________.

Pre-Lab Exercises for PREPARATION AND REACTIONS OF GRIGNARD REAGENTS.
Part D. PREPARATION OF 2-METHYL-3-HEPTANOL (Section 19.2)

NAME (print): ____________________ DATE: __________

INSTRUCTOR: ____________________ LABORATORY SECTION: __________

1. What is the limiting reagent in this procedure? Show your work.

2. Why do you think the particular reagent specified in Exercise 1 was made limiting?

3. Why is it recommended that 2-methylpropanal be freshly distilled?

4. Why must the aldehyde and the diethyl ether in which it is dissolved be anhydrous?

5. Why is the solution of aldehyde added to the Grignard reagent in a dropwise fashion rather than added all at once?

6. What is wrong with storing ethereal solutions in the laboratory bench from one period to the next?

7. Why is technical rather than anhydrous diethyl ether used in the work-up procedure?

8. Why is hydrolysis of the reaction mixture performed with aqueous sulfuric acid that is cold rather than at room temperature?

9. How is unchanged aldehyde removed from the desired product by the work-up procedure used?

10. What happens to any Grignard reagent that remains in the reaction mixture after addition of the aldehyde?

11. The flash point (°C) of 2-methylpropanal is________; its oral LD_{50} (mg/kg) in rats is ________.

12. MSDS information is presently not available for 2-methyl-3-heptanol. In lieu of such data, provide the corresponding information for the analogous compound, 1-heptanol.

NAME (print): ______________________________ DATE: ____________

INSTRUCTOR: ______________________________ LABORATORY SECTION: __________

1. Mechanistically, the reaction to be performed (underline all answers that apply) is (a) an elimination reaction, (b) a nucleophilic substitution reaction, (c) a nucleophilic addition-elimination reaction, (d) an electrophilic substitution reaction, (e) a free-radical substitution reaction, (f) an oxidation of the aromatic substrate, (g) a reduction of the aromatic substrate, (h) none of these.

2. Determine the limiting reagent in this reaction. Show your work.

3. Why is the reduction of nitrobenzene more efficient with tin powder that is free of surface oxides?

4. What changes in the reaction mixture do you expect to see as the reduction progresses?

5. Determine whether adding the specified amount of 12 *M* aqueous sodium hydroxide to the reaction mixture prior to steam distillation is sufficient to make the mixture basic. Show your calculation.

6. How is aniline to be separated from residual nitrobenzene in this procedure?

7. Why is the reaction mixture to be cooled after being saturated with salt but before extraction with diethyl ether?

8. What color is pure aniline? Why are undistilled samples of it often brown in color?

9. The oral LD_{50} (mg/kg) in rats of nitrobenzene and aniline are ________, and ________, respectively.

10. Is aniline listed as a potential carcinogen?

Pre-Lab Exercises for SYNTHESIS OF SULFANILAMIDE. Part B. PREPARATION OF ACETANILIDE (Section 20.2)

NAME (print): ______________________________ DATE: ____________

INSTRUCTOR: ______________________________ LABORATORY SECTION: __________

1. Mechanistically, the reaction to be performed (underline all answers that apply) is (a) an elimination reaction, (b) a nucleophilic substitution reaction, (c) a nucleophilic addition-elimination reaction, (d) an electrophilic substitution reaction, (e) a free-radical substitution reaction, (f) an oxidation of the aromatic substrate, (g) a reduction of the aromatic substrate, (h) none of these.

2. What is the limiting reagent in this reaction? Show your work.

3. How is unchanged aniline to be separated from acetanilide in this procedure?

4. Suppose you plan to make up the specified aqueous solution of sodium acetate *after* you combine acetic anhydride with the aqueous solution of aniline. Why is this not an appropriate strategy?

5. Why is aqueous hydrochloric acid to be combined with aniline in this procedure?

6. Why is concentrated hydrochloric acid to be added *to* water rather than the reverse when you prepare the dilute acid to be used in this procedure?

7. What is the role of the sodium acetate in this reaction?

8. What strategy is to be used if the isolated acetanilide is colored?

9. Is aniline listed as a potential carcinogen?

10. List the toxic fumes evolved in fires involving aniline, acetic anhydride, and acetanilide.

11. The oral LD_{50} (mg/kg) in rats of aniline and acetic anhydride are ________ and________, respectively.

Pre-Lab Exercises for SYNTHESIS OF SULFANILAMIDE. Part C. PREPARATION OF 4-ACETAMIDOBENZENESULFONYL CHLORIDE (Section 20.2)

NAME (print): ______________________________ DATE: __________

INSTRUCTOR: ______________________ LABORATORY SECTION: ________

1. Mechanistically, the reaction to be performed (underline all answers that apply) is (a) an elimination reaction, (b) a nucleophilic substitution reaction, (c) a nucleophilic addition-elimination reaction, (d) an electrophilic substitution reaction, (e) a free-radical substitution reaction, (f) an oxidation of the aromatic substrate, (g) a reduction of the aromatic substrate, (h) none of these.

2. What is the limiting reagent in this reaction? Show your work.

3. Why is it important that the acetanilide to be used in this experiment be dry?

4. Why is ice water rather than warm water to be used for hydrolysis of the reaction mixture?

5. Given the work-up procedure that is to be used, could residual acetanilide contaminate the desired product in its crude state? Explain.

7. Why is it important to break up any lumps of the crude isolated product?

7. Why should the 4-acetamidobenzenesulfonyl chloride be combined with ammonia immediately rather than in the next laboratory period?

8. List the toxic fumes evolved in fires involving chlorosulfonic acid and acetanilide.

9. How are you to destroy residual amounts of chlorosulfonic acid?

10. Why is water *not* an appropriate medium for extinguishing fires involving chlorosulfonic acid?

11. An MSDS is currently not available for *p*-acetamidobenzenesulfonyl chloride. In lieu of this information, provide the following information for the analogous compound, benzenesulfonyl chloride:

a. Extinguishing media are

b. The oral LD_{50} (mg/kg) in rats for the compound is ________.

c. Hazardous combustion products are

Pre-Lab Exercises for SYNTHESIS OF SULFANILAMIDE. Part D. PREPARATION OF 4-ACETAMIDOBENZENESULFANILAMIDE (Section 20.2)

NAME (print): ______________________________ DATE: ____________

INSTRUCTOR: ______________________________ LABORATORY SECTION: __________

1. Mechanistically, the reaction to be performed (underline all answers that apply) is (a) an elimination reaction, (b) a nucleophilic substitution reaction, (c) a nucleophilic addition-elimination reaction, (d) an electrophilic substitution reaction, (e) a free-radical substitution reaction, (f) an oxidation of the aromatic substrate, (g) a reduction of the aromatic substrate, (h) none of these.

2. What is the limiting reagent in this reaction? Show your work.

3. Why will reaction of 4-acetamidobenzenesulfonyl chloride with ammonium hydroxide give the sulfonamide rather than the sulfonic acid?

4. What would account for the exothermicity that might occur when ammonium hydroxide is first added to the crude 4-acetamidobenzenesulfonyl chloride?

5. Why is 6 *M* sulfuric acid to be added after reaction of the sulfonyl chloride with ammonium hydroxide?

6. Why might there be a preference for the use of solid sodium carbonate instead of solid sodium hydroxide for basifying the acidic hydrolysis solution to be obtained in this experiment?

7. Provide two other names under which 4-acetamidobenzenesulfonyl chloride might be listed.

8. List some possible physiological effects of 4-acetamidobenzenesulfonyl chloride.

Pre-Lab Exercises for SYNTHESIS OF SULFANILAMIDE. Part E. PREPARATION OF SULFANILAMIDE (Section 20.2)

NAME (print): ______________________________ DATE: __________

INSTRUCTOR: ______________________________ LABORATORY SECTION: __________

1. Mechanistically, the reaction to be performed (underline all answers that apply) is (a) an elimination reaction, (b) a nucleophilic substitution reaction, (c) a nucleophilic addition-elimination reaction, (d) an electrophilic substitution reaction, (e) a free-radical substitution reaction, (f) an oxidation of the aromatic substrate, (g) a reduction of the aromatic substrate, (h) none of these.

2. Assume you are to prepare 30 mL of *dilute* hydrochloric acid as directed in the experimental procedure. Detail how would you do this and specify the molarity of the solution that results.

3. What is the role of hydrochloric acid in this reaction?

4. Assume 30 mL of dilute HCl is used in this procedure. How much sodium carbonate would be required to neutralize this much acid?

5. Is sulfanilamide listed as a potential carcinogen?

6. List the target organs for sulfanilamide.

7. The oral LD_{50} (mg/kg) in rats of sulfanilamide is ________.

8. List the toxic fumes evolved in fires involving sulfanilamide.

Pre-Lab Exercises for SYNTHESIS OF 1-BROMO-3-CHLORO-5-IODOBENZENE. Part B. PREPARATION OF 4-BROMOACETANILIDE (Section 20.3)

NAME (print): ______________________ DATE: __________

INSTRUCTOR: ______________________ LABORATORY SECTION: __________

1. Mechanistically, the reaction to be performed (underline all answers that apply) is (a) an elimination reaction, (b) a nucleophilic substitution reaction, (c) a nucleophilic addition-elimination reaction, (d) an electrophilic substitution reaction, (e) a free-radical substitution reaction, (f) an oxidation of the aromatic substrate, (g) a reduction of the aromatic substrate, (h) none of these.

2. Write the structures of the two other monobromo products that might be formed in this reaction and *circle* the one *least* likely to be formed in the reaction.

3. What is the limiting reagent in this reaction? Show your work.

4. If the crude product appears colored, how is it to be decolorized?

5. What do you expect to see when ice-cold water is added to the reaction mixture?

6. What should you do if glacial acetic acid comes in contact with your skin.

7. Why should glassware containing residues of bromine *not* be rinsed with acetone?

8. The flash point (°C) of glacial acetic acid is ________.

9. Is glacial acetic acid listed as a potential carcinogen?

10. Is an MSDS available for 4-bromoacetanilide?

Pre-Lab Exercises for SYNTHESIS OF 1-BROMO-3-CHLORO-5-IODOBENZENE. Part C.
PREPARATION OF 4-BROMO-2-CHLOROACETANILIDE (Section 20.3)

NAME (print): ______________________________ DATE: __________

INSTRUCTOR: ______________________________ LABORATORY SECTION: __________

1. Mechanistically, the reaction to be performed (underline all answers that apply) is (a) an elimination reaction, (b) a nucleophilic substitution reaction, (c) a nucleophilic addition-elimination reaction, (d) an electrophilic substitution reaction, (e) a free-radical substitution reaction, (f) an oxidation of the aromatic substrate, (g) a reduction of the aromatic substrate, (h) none of these.

2. Write the balanced equation for formation of molecular chlorine from sodium chlorate and hydrochloric acid.

3. What is the limiting reagent in this reaction? Show your work.

4. Why is a gas trap to be used in this experiment if the reaction cannot be performed in a hood?

5. What do you expect to observe in the reaction vessel as the chlorination proceeds?

6. Should recrystallization of the product be necessary, what solvent is to be used?

7. What technique is to be used to destroy chlorine if it is present in any of the filtrates or washes obtained in this experiment?

8. What is the solubility of sodium chlorate in water?

9. The oral LD_{50} (mg/kg) in rats of sodium chlorate is ________.

10. Underline the media appropriate for extinguishing fires involving sodium chlorate: Water Carbon dioxide Chemical powder Foam

Pre-Lab Exercises for SYNTHESIS OF 1-BROMO-3-CHLORO-5-IODOBENZENE. Part D. PREPARATION OF 4-BROMO-2-CHLOROANILINE (Section 20.3)

NAME (print): ______________________ DATE: __________

INSTRUCTOR: ______________________ LABORATORY SECTION: __________

1. Mechanistically, the reaction to be performed (underline all answers that apply) is (a) an elimination reaction, (b) a nucleophilic substitution reaction, (c) a nucleophilic addition-elimination reaction, (d) an electrophilic substitution reaction, (e) a free-radical substitution reaction, (f) an oxidation of the aromatic substrate, (g) a reduction of the aromatic substrate, (h) none of these.

2. What is the limiting reagent in this reaction? Show your work.

3. What is the role of hydrochloric acid in promoting the hydrolysis reaction?

4. Determine whether the specified amount of aqueous sodium hydroxide that is to be used is sufficient to make the reaction mixture basic. Show your work.

5. Why is *flameless* heating specified for heating the reaction mixture under reflux?

6. What change in the originally homogeneous solution may you observe as the reaction proceeds?

7. If you are to recrystallize the crude product, what strategy is to be used if oiling out rather than crystallization occurs?

8. The flash point (°C) of ethanol is ________.

9. The oral LD_{50} (mg/kg) in rats of ethanol is ________.

Pre-Lab Exercises for SYNTHESIS OF 1-BROMO-3-CHLORO-5-IODOBENZENE. Part E. PREPARATION OF 4-BROMO-2-CHLORO-6-IODOANILINE (Section 20.3)

NAME (print): ______________________________ DATE: __________

INSTRUCTOR: ______________________________ LABORATORY SECTION: __________

1. Mechanistically, the reaction to be performed (underline all answers that apply) is (a) an elimination reaction, (b) a nucleophilic substitution reaction, (c) a nucleophilic addition-elimination reaction, (d) an electrophilic substitution reaction, (e) a free-radical substitution reaction, (f) an oxidation of the aromatic substrate, (g) a reduction of the aromatic substrate, (h) none of these.

2. What is the limiting reagent in this reaction? Show your work.

3. Why does iodination rather than chlorination occur when 4-bromo-2-chloroaniline is treated with iodine monochloride?

4. Why is the iodination being performed on 4-bromo-2-chloroaniline rather than 4-bromo-2-chloroacetanilide?

5. What is a target organ for iodine monochloride in the human body?

6. The melting point of iodine monochloride is________; its boiling point is ________.

7. Underline the media appropriate for extinguishing fires involving sodium iodine monochloride: Water Carbon dioxide Chemical powder Foam

Pre-Lab Exercises for SYNTHESIS OF 1-BROMO-3-CHLORO-5-IODOBENZENE. Part F. PREPARATION OF 1-BROMO-3-CHLORO-5-IODOBENZENE (Section 20.3)

NAME (print): ______________________________ DATE: __________

INSTRUCTOR: ______________________________ LABORATORY SECTION: __________

1. Mechanistically, the reaction to be performed (underline all answers that apply) is (a) an elimination reaction, (b) an electrophilic substitution reaction, (c) an SN_2 reaction, (d) an oxidation of the aromatic substrate, (e) a reduction of the aromatic substrate, (f) none of these.

2. What is the limiting reagent in this reaction? Show your work.

3. List the target organs for sodium nitrite.

4. The oral LDL_0 (mg/kg) in humans of sodium nitrite is _______.

PL. 148

NAME (print): ______________________________ DATE: __________

INSTRUCTOR: ______________________ LABORATORY SECTION: ________

1. Write an equation for the reaction of *tert*-butylcatechol with sodium hydroxide solution that is responsible for removing the inhibitor from commercial styrene and explain how the extraction accomplishes the desired separation.

2. Write formulas for the products of the *disproportionation* reaction between two styryl radicals, $CH(C_6H_5)CH_3$.

3. What determines whether or not decantation rather than filtration may be used to separate a solid from a liquid?

4. Indicate whether the following statement is true (T) or false (F): Styrene and *tert*-butyl peroxybenzoate are both listed as potential tumorigenics.

____

5. The LCL_{50} (ppm) for inhalation of styrene by humans is _______ during _______ min.

6. The flash points (°C) of styrene, *tert*-butyl peroxybenzoate, and xylene are _______, _______, and _______.

Pre-Lab Exercises for PREPARATION OF NYLON-6,10
(Section 21.3)

NAME (print): ______________________ DATE: __________

INSTRUCTOR: ______________________ LABORATORY SECTION: __________

1. Why is decanedioyl chloride rather than decanedoic acid used in this experiment?

2. Explain how reaction between decanedioyl chloride in dichloromethane solution and 1,6-hexanediamine in aqueous solution occurs although the two immiscible solutions are not mixed.

3. Why should the tip of the separatory funnel containing the 1,6-hexanediamine solution be placed no more than 1 cm above the surface of the dichloromethane solution when making the addition?

4. Why is the aqueous solution added to the dichloromethane solution, rather than *vice versa*?

5. Note that decanedioyl chloride and 1,6-hexanediamine are used in equimolar amounts, whereas in many experiments one reactant is used in considerable molar excess. Why is the equimolar ratio used here?

6. Why is the formic acid solution of the polymer evaporated in the hood rather than at the laboratory bench?

7. The flash point (°C) of decanedioyl dichloride is ________.

8. List the toxic fumes that are evolved in fires involving dichloromethane, 1,6-hexanediamine and decanedioyl chloride.

9. The oral LD_{50} (mg/kg) in rats of 1,6-hexanediamine is ________, whereas that of formic acid is ________.

NAME (print): ______________________________ DATE: __________

INSTRUCTOR: ______________________________ LABORATORY SECTION: __________

1. Why is the mixture of sugars derived from hydrolysis of sucrose commonly referred to as "invert" sugar?

2. What effect would the presence of air bubbles in the polarimeter tube have on the value of the measured rotation?

3. What effect would incomplete transfer of the hydrolysis mixture from the reaction flask to the volumetric flask have on the value of the measured rotation?

4. The oral LD_{50} (mg/kg) in rats of sucrose is ________.

5. Indicate whether the following statement is true (T) or false (F): Sucrose is a reducing sugar.

   ________.

6. What is the role of hydrochloric acid in the hydrolysis of sucrose?

PL. 154

Pre-Lab Exercises for CLASSIFICATION TESTS FOR CARBOHYDRATES (Section 22.4)

NAME (print): ______________________________ DATE: __________

INSTRUCTOR: ______________________________ LABORATORY SECTION: __________

1. Indicate whether the following statement is true (T) or false (F): Tollens' and Benedict's tests depend on the ability of carbohydrates to oxidize the test reagent. ________.

2. Explain why nonreducing sugars give a positive Barfoed's test more slowly than do reducing sugars.

3. What is likely to happen if Tollens' test is performed in a test tube that is not clean?

4. What do you expect to see in the case of a positive Benedict's test?

5. What do you expect to see in the case of a positive Barfoed's test?

6. Why are both glucose and sucrose recommended as carbohydrates to test for purposes of comparison when performing Benedict's test on an unknown?

7. Why is it important not to heat the test solution for longer than 5 minutes when performing Barfoed's test?

8. What purpose is served by citrate ion in the stock solution of Benedict's reagent?

NAME (print): ______________________________ DATE: __________

INSTRUCTOR: __________________________ LABORATORY SECTION: ________

1. What is the function of saturated aqueous sodium bisulfite solution in this procedure?

2. Would you expect osazone formation to occur faster or slower in strongly acidic media as compared to weakly acidic media? Explain your answer.

3. Why does reaction of phenylhydrazine with aldoses and ketoses stop after the first two carbon atoms in the chain have been converted to hydrazone functions?

4. Why is sodium acetate required if phenylhydrazine hydrochloride is used as the source of phenylhydrazine?

5. Why would D-glucose and D-mannose be expected to form the same osazone upon reaction with phenylhydrazine?

6. Indicate whether the following statement is true (T) or false (F): Phenylhydrazine and its hydrochloride are both listed as a potentially carcinogenic compounds.

____

7. List the toxic fumes evolved upon combustion of phenylhydrazine and its hydrochloride.

8. Underline the media that are appropriate for extinguishing fires involving glucose, fructose, and phenylhydrazine hydrochloride:

Water Carbon dioxide Chemical powder Foam